45803

D0081070

CLARK STATE COMMUNITY COLLEGE LIBRARY

Advanced Circuit Analysis

PAUL E. BENNETT

University of Southern Indiana

SAUNDERS
HBJ

Saunders College Publishing

A Harcourt Brace Jovanovich College Publisher

Forth Worth   Philadelphia   San Diego   New York   Orlando   Austin   San Antonio
Toronto   Montreal   London   Sydney   Tokyo

OCLC # 2450 3732

TK
454
.B3998
1992

To the late Dr. Carl S. Roys, who I considered to be my best teacher as an undergraduate. His emphasis on the basic physics of electrical circuits and devices is reflected in this text.

Copyright © 1992 by Harcourt Brace Jovanovich, Inc.

All rights reserved. No part of this publication may be reproduced or transmitted in any form or by any means, electronic or mechanical, including photocopy, recording or any information storage and retrieval system, without permission in writing from the publisher.

Requests for permission to make copies of any part of the work should be mailed to: Permissions Department, Harcourt Brace Jovanovich, Inc., 8th Floor, Orlando, Florida 32887.

ISBN: 0-15-501843-4

Library of Congress Catalog Card Number: 91-30390

Printed in the United States of America

Contents

PREFACE   vii

III

Preface

Introductory courses in engineering technology circuit analysis do not ordinarily provide coverage of some important aspects of the subject because students rarely have any background in calculus. This textbook does provide coverage of these subjects. The specific subjects include transient and steady-state circuit analysis in the complex frequency or s-plane. And, although the emphasis is on passive circuits, controlled sources are discussed so that students will also be able to work with active circuits. No important concept is ignored.

Students who use this book should be able to differentiate and integrate simple functions such as sinusoids, exponentials, ramps, step functions and products of some of these. When more than one mathematical method is available for a particular task, the method chosen is one that should appeal to students not especially interested in mathematical techniques. Technology students are more likely to be able to follow a text where functions are illustrated by curves as well as by formulas. When appropriate, the discussions in this book are supported by references to the physics of the component or circuit.

The object of this book is to provide a reasonably complete set of tools to use in the analysis of circuits. The examples used illustrate both methods of solution and certain important aspects of circuit behavior. The examples and problems are similar to those that might be found in real circuits.

The appendixes include an introduction to SPICE, a program that can be used to perform both transient and steady-state circuit analyses. Students often receive one version of SPICE free of charge, but little documentation provided with it. This appendix should fill in some gaps. Because many of the problems provided can be solved by either computer or manual methods, no special problems have been provided. Both methods may be used on a given problem with no loss of instructional value. The use of manual methods provides a good introduction to the general concepts of circuit analysis, but of course the computer will be used in real-life situations, especially for more complex circuits.

Chapter 1 provides a review of basic circuit analysis. The techniques discussed are vitally important in the application of advanced methods of analysis. Although the student may already have studied DC and AC circuits, the review should prove useful. Aspects of basic electrical theory that particularly apply to the material in later chapters are emphasized so students get a good base knowledge of the physical aspects of circuit theory.

Chapter 2 includes a discussion of waveforms, a topic especially important in transient applications.

Chapter 3 provides an introduction to the differential equations of electric circuits. There is coverage of forced and natural responses, initial conditions, and the final responses of circuits. For those interested in the analysis of electromechanical devices, there is a short discussion of transients in mechanical devices.

Chapter 4 introduces the Laplace transform, and tables of transform pairs and operations are developed. These tables are short, but sufficiently complete so that all problems can be solved with them. If students need a larger table, a number are readily available. However, the use of computers for solutions will probably make more complete tables unnecessary. A brief introduction to the direct solution of differential equations is given here, and there is a section on partial fraction expansion. The initial and final value theorems are introduced in this chapter. They will be of particular value to students who go on to study automatic control systems.

Chapter 5 deals with the solution of transient problems in electric circuits—the most important topic in the book. Concepts covered include the s-plane representation of sources, linear components, and initial conditions. Some of the examples were chosen to illustrate the responses of specific circuits as well as to demonstrate the method of solution.

Chapter 6 covers steady-state circuit analysis. It includes descriptions of transfer functions and pole-zero plots. This should provide further grounding in s-plane concepts and solidify understanding of the complex frequency plane. Bode plots are also covered here.

An appendix on the use of SPICE covers both transient solutions and frequency responses. While it doesn't provide a complete explanation for all possible types of circuit problems, it is complete enough for solving problems found in this book. Other appendixes cover Cramer's rule, the development of some Laplace transforms, and the development of a quick method of solution for transients that have sinusoidal components in their responses.

I would like to acknowledge the expert guidance of Phillip Menzies, the editor during the major part of the formation of this text. I am grateful to the following people who were kind enough to read some or all of the manuscript: Edward J. Milano, University of Hartford; Jerome Padak, Bailey Technical School; W. Banzhaf; Russell E. Puckett, Texas A & M University; Charles Roby, Clemson University; James L. Hales, University of Pittsburgh at Johnstown; John Slough, DeVry Institute of Technology; and Guillermo Rico, New Mexico State University. I would also like to express my appreciation to my wife Carolyn, both for her patience with the demands on my spare time and her valuable suggestions on writing style.

CHAPTER 1

Basic Circuit Analysis

1.1 OBJECTIVES

On completion of this chapter, you should be able to:

- Define the basic variables involved in electric circuit analysis, including current, voltage, energy, and power.
- Discuss the characteristics of the basic linear electric circuit elements—resistance, capacitance, and both self- and mutual inductance.
- Describe the characteristics of dependent and independent current and voltage sources.
- Apply the techniques used in the analysis of complex circuits, including both mesh and nodal analysis, the principle of superposition, and Thévenin's and Norton's theorems.

1.2 INTRODUCTION

Many of the circuit analysis concepts to be described here are covered in DC and AC circuit analysis courses. However, the significance of certain features of the theory may not have been obvious when the emphasis was on steady-state analysis. This is particularly true if the theory was covered before the student had any knowledge of calculus.

For this reason, and because the basic procedures are important in transient circuit analysis, they are briefly reviewed in this chapter. Along with these procedures, several techniques are discussed that should aid in the solution of practical problems.

1.3 VARIABLES IN ELECTRIC CIRCUITS

Four fundamental measurable variables are involved in the analysis of electric circuits: energy, current, voltage, and power. Circuit analysis usually involves the calculation of one or several of these.

1.3.1 Energy

In commercial applications, electricity was originally generated to perform one major function: the efficient transfer of energy from one place to another, in order to provide illumination in locations remote from the generating station. This purpose still holds today even in electronic applications, where the results achieved depend on the transmission of small amounts of signal energy in order to transfer information.

Energy is defined as the capacity to do work. When work is done, energy is used up. The SI unit for energy is the joule (J). It takes 1 joule to raise the temperature of 1 cubic centimeter of water 1 kelvin. In electrical systems, if 1 volt causes 1 ampere to flow, the energy expended in 1 second is 1 watt-second, or 1 joule.

1.3.2 Current and Kirchhoff's Current Law

Current is defined as the rate of flow of electric charge in the form of electrically charged particles. The individual particles are usually the negatively charged electrons. In solid conductors, the particles that can be moved a significant distance are electrons. Current flow in semiconductors also involves the motion of electrons. In liquid electrolytes, such as the sulfuric acid solution in storage batteries, both electrons and positively charged ions flow in opposite directions.

Current is measured in amperes (A). One ampere is defined as the flow of 1 coulomb of electric charge per second. This definition is expressed in the equation

$$i = \frac{dq}{dt} \text{ A} \tag{1-1}$$

where i is current in amperes, q is electric charge in coulombs, and t is time in seconds.

There is a repulsion between like charges and an attraction between unlike charges. As a result of the repulsion, any electrons entering a junction or node of a circuit will cause an equal number of electrons to leave. Kirchhoff's current law expresses this relationship as an equation. It states that the algebraic sum of currents entering and leaving a node in a circuit is zero. A node can be a junction, or it can be a section of a circuit containing either sources or impedances or both. Kirchhoff's current law can be expressed as

$$\Sigma I_n = 0 \text{ A} \tag{1-2}$$

In this form of the equation, currents entering and leaving the node are assigned opposite signs.

EXAMPLE 1–1   Figure 1–1 illustrates the application of Kirchhoff's current law at a junction. The dashed lines indicate connections to other parts of the circuit. Here,

$$I_1 + I_2 - I_3 + I_4 = 0 \text{ A} \tag{1-3}$$

We have given positive signs to currents flowing into the node and negative signs to currents flowing out. The basic concept of Kirchhoff's current law is that the total charge flowing into any junction or part of a circuit is matched by an equal charge flowing out.

   If the values of currents I_1, I_2, and I_3 are 1, 5, and 2 A, as shown in Figure 1–1, we can substitute these values into Equation 1–3 to solve for I_4, obtaining

$$1 + 5 - 2 + I_4 = 0$$

or

$$I_4 = -4 \text{ A}$$

As previously mentioned, the negative sign shows that I_4 is flowing out of the junction.

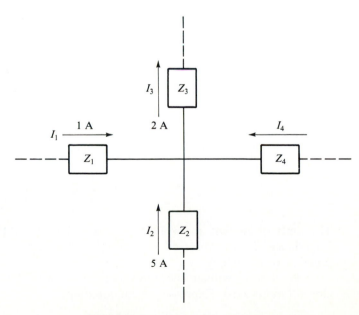

FIGURE 1–1   Kirchhoff's current law

1.3.3 Voltage, Ohm's Law, and Kirchhoff's Voltage Law

What causes charged particles to flow? In flowing through a solid conductor at normal temperatures, they will encounter opposition in the form of collisions with particles that are not free to flow. Some force is required to overcome this opposition. We call this force *potential difference*, or *voltage*. The basic unit of potential is the volt (V).

A volt is defined as the potential difference between two points such that it takes 1 joule of energy to move a charge of 1 coulomb from one of the points to the other against the electrostatic force produced by the potential. The equation describing this relationship is

$$V = \frac{W}{Q} \text{ V} \qquad (1\text{--}4)$$

where W is energy in joules and Q is charge in coulombs. The units of potential are joules per coulomb, or volts. Note that, despite its being defined on the basis of energy, the volt is not a unit of energy.

If the voltage is not constant, the more general form of the equation should be used:

$$v = \frac{dw}{dq} \text{ V} \qquad (1\text{--}5)$$

That is, the (instantaneous) voltage is equal to the rate of change of energy with respect to charge.

The preceding discussion implies that voltage is the force that causes current to flow through an electric circuit. A basic principle of physics,

$$\text{effect} = \frac{\text{cause}}{\text{opposition}}$$

is involved. In the case of a circuit, the cause is the potential difference across the circuit and the effect is the current that flows. Because the opposition impedes the flow of current, it is called the circuit *impedance*. In electrical terms, we write Ohm's law as

$$I = \frac{V}{Z} \text{ A} \qquad (1\text{--}6)$$

The units of current and voltage in this equation are the ampere and the volt. Impedance (Z) is expressed in ohms (Ω). For some circuit components, it is a function of frequency.

Kirchhoff's voltage law states that the algebraic sum of the voltages around a closed loop is zero. Expressed as an equation,

$$\Sigma V_i = 0 \text{ V} \qquad (1\text{--}7)$$

The law is illustrated in Figure 1–2. The elements in the circuit can be voltage sources or impedances. In the circuit shown, the polarities of the elements' voltages do not necessarily indicate whether the element is a source or an impedance. The sign of the drop across an impedance will depend on the direction of current flow. It is possible for voltage sources to aid or oppose the flow of current, depending on the way they are connected in the circuit.

Summing up the voltages around the loop, we find that

$$V_1 + V_2 + V_3 + V_4 + V_5 = 0 \text{ V} \tag{1–8}$$

If we go around the loop in the direction of current flow, a voltage rise across an element can be considered to have a positive sign and a voltage drop to have a negative sign.

A nonzero sum of the voltages around the loop in Figure 1–2 would mean that the last terminal of element Z_5 would not be at the same potential as the first terminal of element Z_1. These terminals are connected by a conductor that we assume to have zero resistance. Ohm's law can be used to find the current that would flow in the wire due to an assumed voltage E. The current would be

$$I = \frac{E}{Z} = \frac{E}{0} = \infty \text{ A} \tag{1–9}$$

This is an impossible result, of course.

EXAMPLE 1–2 The magnitudes of voltages V_1 through V_4 are given in Figure 1–2. Find the value of V_5.

By Equation 1–7,

$$\Sigma V_i = 0 \text{ V}$$

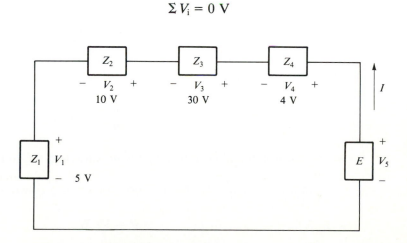

FIGURE 1–2 Kirchhoff's voltage law

Summing all voltages, going clockwise around the loop, we obtain

$$5 + 10 + 30 + 4 + V_5 = 0$$

or

$$V_5 = 49 \text{ V}$$

1.3.4   Power

Power is defined as the rate of producing energy or doing work. The SI unit of power is the watt (W). A watt is defined as 1 joule per second. If both sides of Equation 1–1, which defines current, are integrated, we can solve for the charge:

$$q = \int i\, dt \text{ C} \tag{1–10}$$

Differentiating both sides of Equation 1–10, we find that

$$dq = i\, dt \tag{1–11}$$

Substituting Equation 1–11 into the energy equation (Equation 1–5), solving for dw, and integrating, we obtain

$$w = \int vi\, dt \text{ J} \tag{1–12}$$

Differentiating both sides of Equation 1–12, we see that the rate of production or use of energy is

$$\frac{dw}{dt} = vi \text{ W} \tag{1–13}$$

Thus, the rate of change of energy, or power, is

$$p = \frac{dw}{dt} = vi \text{ W} \tag{1–14}$$

where the instantaneous power (p) is in watts and the other units are as defined earlier.

In a DC circuit, the instantaneous power is a constant given by

$$P = VI \text{ W} \tag{1–15}$$

Since P is a constant, the average power in a DC circuit is also equal to VI.

In an AC circuit, where the excitation is sinusoidal, the instantaneous voltage is expressed by

$$v = E_m \sin \omega t \ \text{V} \tag{1-16}$$

where E_m is the peak value of the voltage and ω is the frequency. The current will then be sinusoidal with a phase shift $\theta°$ with respect to the voltage. Mathematically,

$$i = I_m \sin(\omega t + \theta) \ \text{A} \tag{1-17}$$

where I_m is the peak value of the current and, again, ω is the frequency. The instantaneous power is the product vi:

$$p = vi = E_m I_m (\sin \omega t)[\sin(\omega t + \theta)] \ \text{W} \tag{1-18}$$

Using several trigonometric identities, we find that

$$p = \frac{E_m I_m}{2} [(\cos \theta)(1 - \cos 2\omega t) + (\sin \theta)(\sin 2\omega t)] \ \text{W} \tag{1-19}$$

or

$$p = \frac{E_m I_m}{2} \cos \theta - \frac{E_m I_m}{2} \cos \theta \cos 2\omega t + \frac{E_m I_m}{2} \sin \theta \sin 2\omega t \ \text{W} \tag{1-20}$$

The first term is a constant, and the other two are sinusoidal. The sinusoidal components have a zero time average; thus, the average power is equal to

$$P = \frac{E_m I_m}{2} \cos \theta \ \text{W} \tag{1-21}$$

Definitions have been developed for AC volts and amperes that provide the same power in resistive circuits as the same number of DC volts and amperes would. These are called *effective*, or *root-mean-square* (rms), values. For sinusoidal waveforms, the rms values can be found by dividing the peak values of voltage and current by $\sqrt{2}$.

The average power is

$$P = EI \cos \theta \ \text{W} \tag{1-22}$$

when the voltage and current are measured in rms values. The product of voltage and current is called the *apparent power*:

$$P_A = EI \ \text{VA} \tag{1-23}$$

Here, the voltage and current are also in rms units.

1.4 SOURCES OF POWER

Sources of electrical power for circuits have a number of forms. Some provide reasonably constant output voltages or currents. Some provide outputs whose amplitudes are proportional to a current or voltage at some other point, either inside or outside the circuit. In most cases, these arise as a part of the equivalent circuit of amplifying devices.

Most sources have a constant DC or sinusoidal AC output, but other output waveshapes are possible, such as rectangular pulses, ramps, or exponentials. These can be single or repetitive waveforms. On occasion, complex waveforms are encountered, often as a result of nonlinearity.

1.4.1 Real and Ideal Sources

Ideal voltage sources produce a constant-amplitude voltage regardless of the current supplied to the load. Similarly, ideal current sources furnish a constant-amplitude current to any load. Ideal sources can be approximated using real sources with electronic circuits added that compensate for the changes in output voltage caused by load changes. We call these sources *regulated power supplies*.

The compensation in regulated sources is not complete, because there is still a small change in output with load changes. Also, the compensation is effective only over a limited range of load currents. In many cases, however, such regulated sources are close enough to the ideal sources that we may assume that they are ideal.

All practical electrical sources are what we call *real sources*. The voltage supplied by a real voltage source decreases as the current drawn from the source increases. The amount of drop is directly proportional to the current. This indicates that the equivalent circuit consists of an ideal voltage source in series with a resistance, as shown in Figure 1–3(a).

In the figure, E_S is the voltage appearing at the terminals when Z_L is infinite. R_S, the source resistance, is the value of internal resistance that would cause full-load voltage to appear at the terminals when full-load current flows. Application of Kirchhoff's voltage law shows that the voltage across R_S is the difference between the no-load voltage E_S and the full-load terminal voltage v_{FL}:

$$v_R = E_S - v_{FL}$$

The current from a real current source decreases as the load resistance increases. The simplest equivalent circuit that would provide this characteristic is an ideal current source in parallel with a resistor, as shown in Figure 1–3(b). Here, the ideal current source supplies a current I_S equal to the current supplied to a short circuit across the load terminals. The current I_S will have two parts, one part going through the source resistance and the rest through the load impedance.

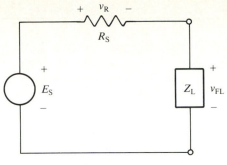

(a) Real voltage source

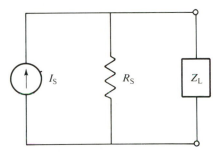

(b) Real current source   **FIGURE 1–3**   *Real sources*

1.4.2   Independent and Dependent Sources

Independent real sources provide a constant-amplitude output voltage or current at all times, under normal operating conditions. The equivalent circuits for amplifying devices, such as transistors, include dependent sources to account for the gain of the device. Dependent sources are sources whose magnitudes depend on some other voltage or current, usually in a part of the circuit that is isolated from the source. The presence of dependent sources may cause difficulty in, but does not prevent, analysis of the circuits they are found in.

Different symbols are often used to represent dependent versus independent sources. This scheme is justified by the fact that the two types of sources are fundamentally different. Figure 1–4 shows the symbols used for various types of controlled sources. Both dependent voltage sources and dependent current sources can be controlled by either current or voltage. Figure 1–4(a) is of a voltage-controlled voltage source (VCVS). Figure 1–4(b) shows a current-controlled current source (CCCS). A voltage-controlled current source (VCCS) is illustrated in Figure 1–4(c), and Figure 1–4(d) shows a current-controlled voltage source (CCVS).

If the multiplying factor k is large enough, these dependent sources can control large currents or voltages by small currents or voltages. It is because of this feature that they can represent amplification in a transistor equivalent circuit.

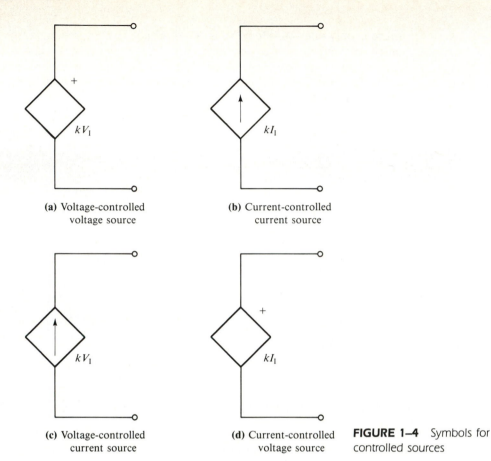

(a) Voltage-controlled
voltage source

(b) Current-controlled
current source

(c) Voltage-controlled
current source

(d) Current-controlled
voltage source

FIGURE 1–4 Symbols for
controlled sources

Controlled sources have some resemblance to ideal independent sources. The terminal voltage of controlled voltage sources does not necessarily depend on the current through them. Nor does the current through controlled voltage sources usually depend directly on the voltage across them. Controlled sources also resemble impedances in that their voltage or current is not necessarily fixed at some predetermined value.

1.4.3   Phasor Representation of Sinusoidal AC

In AC circuits, the voltages and currents are sinusoidal. For instance, a voltage source may provide the output

$$e(t) = A_{\mathrm{m}} \sin(\omega t + \theta) \text{ V} \qquad (1\text{--}24)$$

where A_{m} is the peak amplitude of the sine wave in volts, ω is the frequency in radians per second, and θ is the phase angle of the wave at $t = 0$ seconds, or what we call the *phase shift*. In general, A_{m}, ω, and θ are variables. In many systems, however, there is only one frequency, so ω is a constant and the remaining variables are A_m and θ. In these single-frequency systems, we can represent the sine wave in a manner that is easily used in circuit calculations if we include both the magnitude and phase shift. This method is called *phasor representation* and can be employed for both current and voltage.

A theoretical basis for phasors will be given in Chapter 2, but in this section we will concentrate on their use. The phasor is like a vector in that it has a magnitude and an angle. The magnitude is the amplitude of the voltage or current. The angle is the phase shift with respect to a reference sine wave. Accordingly, vector operations apply to phasors.

In some steady-state applications, effective or RMS values may be used for phasor magnitudes. In many transient analyses, the response will be the sum or the product of a phase-shifted sinusoid and an exponential. In these cases, the peak value will be easiest to use. As a result, in this text, magnitudes of phasors are given as peak values of the associated sinusoids. By this scheme, the voltage given in Equation (1–24) is, in phasor form,

$$E = A_m \angle \theta \text{ V} \tag{1-25}$$

If the original voltage were given by

$$e(t) = A_m \cos(\omega t + \theta) \text{ V}$$

the equation should be converted to a sine wave before setting it in phasor form. The sine wave will be

$$e(t) = A_m \sin(\omega t + \theta + 90°) \text{ V} \tag{1-26}$$

and the phasor form is

$$E = A_m \angle (\theta + 90°) \text{ V} \tag{1-27}$$

A phase-shifted sine wave can be shown to be the sum of a sine wave and a cosine wave by the trigonometric identity

$$\sin(x + y) = \sin x \cos y + \cos x \sin y$$

Using this identity with the sine wave given in Equation (1–24), we obtain

$$\begin{aligned} e(t) &= A_m \sin(\omega t + \theta) \\ &= A_m \cos \theta \sin \omega t + A_m \sin \theta \cos \omega t \end{aligned} \tag{1-28}$$

The coefficient of the sine wave is $A_m \cos \theta$, and the coefficient of the cosine wave is $A_m \sin \theta$.

1.5 PASSIVE CIRCUIT ELEMENTS _____

The impedances that provide opposition to the flow of current are of three basic types: resistors, capacitors, and inductors. Another device, which is made up of coupled inductors, is the transformer.

1.5.1 Resistors

Most resistances appear as the opposition of a conductor or semiconductor to current flow. When current flows through a resistance, power is dissipated as heat. Ohm's law for resistance is usually written

$$V = IR \text{ V} \tag{1-29}$$

The unit of resistance is the ohm.

Using Ohm's law and Equation (1–15) for power, we can develop two other equations for calculating power. Solving Ohm's law for current yields

$$I = \frac{V}{R} \text{ A} \tag{1-30}$$

and substituting this into Equation (1–15) gives

$$P = I^2 R \text{ W} \tag{1-31}$$

Using Equation (1–30) directly and substituting into the power equation, we get

$$P = \frac{V^2}{R} \text{ W} \tag{1-32}$$

The resistance of a real resistor varies somewhat with frequency, but the change is usually small. At least for narrow frequency ranges and reasonably constant temperatures, it can usually be considered to be a constant.

The total AC resistance includes more than the opposition of the material to the flow of current as observed in DC circuits. Any loss or transfer of power from a circuit will cause an increase in resistance over the DC value.

In AC circuits, power dissipation can be increased by skin effect in wires, hysteresis in ferromagnetic cores, and eddy currents in conductors cut by changing magnetic flux lines. When wavelengths are not very long compared to circuit conductor lengths, power can be transferred out of the circuit by radio waves. Although this power may not all be dissipated, it is lost from the circuit.

The total resistance of a circuit includes both the DC and AC effects and is called *effective resistance*. Ohmmeters cannot measure the AC part of the resistance, because they are DC instruments. The total effective resistance can be determined by measuring power and either voltage or current and using one of the following equations:

$$R_{\text{eff}} = \frac{V^2}{P} \ \Omega \qquad\qquad (1\text{--}33)$$

$$R_{\text{eff}} = \frac{P}{I^2} \ \Omega \qquad\qquad (1\text{--}34)$$

In many circuits, the AC effects on resistance are minimal and can be ignored.

1.5.2 Capacitors

A pure capacitance does not dissipate energy. It can store energy if negative electric charges in the form of free electrons are concentrated on one plate of the capacitor. Since the negative charges repel each other, some energy must be used in placing them near one another. This energy, the stored energy, resembles the potential energy used to raise a mass against the force of gravity.

When negative charges collect on one plate, the resulting electric field causes an equal number of free electrons to be forced away from the other plate, leaving a net positive charge on that plate of the capacitor. As a result, there is an electric field between the plates.

The ability of a capacitor to store energy is determined by its *capacitance*. The total charge stored is directly proportional to both the capacitance and the voltage and is given by

$$q = Cv \ \text{C} \qquad\qquad (1\text{--}35)$$

In this equation, charge is in coulombs and capacitance is in farads. The total charge is also proportional to the time integral of the rate of flow of charge; hence,

$$q = \int_0^t i \, dt \ \text{C} \qquad\qquad (1\text{--}36)$$

Upon integrating, there is a constant of integration which is the charge on the capacitor at the start of the integration interval. Substituting the right side of Equation (1–35) into the left side of Equation (1–36) and dividing by C, we obtain

$$v = \frac{1}{C} \int_0^t i \, dt \ \text{V} \qquad\qquad (1\text{--}37)$$

The constant of integration here is the voltage across the capacitor at $t = 0$ seconds. This voltage represents the energy stored in the capacitor at the start of integration. We can call it the *initial condition for the capacitor*.

If both sides of Equation (1–37) are differentiated with respect to time, we get another equation relating the voltage across a capacitor to the current flowing through it, namely,

$$i = C\frac{dv}{dt} \text{ A} \qquad (1\text{--}38)$$

Equations (1–37) and (1–38) express the relation between voltage and current in capacitors. Either is a counterpart of Ohm's law for resistors.

To give an instantaneous change in voltage, an instantaneous change in charge is required, as can be seen in Equation (1–35). In Equation (1–36), the charge can be seen to be proportional to the time integral of current. It would take an infinite current to cause an instantaneous change in the value of the integral or in the charge. In general, rapid changes in voltage require large currents. As a result, we can say that a capacitor resists fast changes in voltage.

For AC applications, where the voltages and currents are sinusoidal, there is a simpler method of relating voltage and current: through the use of the impedance. If the voltage is sinusoidal, we have

$$v = \text{B} \sin \omega t \text{ V} \qquad (1\text{--}39)$$

Substituting this expression for v into Equation (1–38), we obtain

$$i = C\frac{d(\text{A} \sin \omega t)}{dt} \text{ A} \qquad (1\text{--}40)$$

Performing the differentiation, we find that

$$i = C(\text{A}\omega \cos \omega t) = \omega C[\text{A} \sin(\omega t + 90°)] \text{ A} \qquad (1\text{--}41)$$

That is, the current equals ωC times the voltage waveform shifted 90° in phase. The voltage would then be equal to the current divided by ωC, with a negative phase shift of 90°. The magnitude of the impedance, which is voltage divided by current, is consequently $1/\omega C$. To account for the phase shift, we say that the impedance of a capacitor has a $-90°$ phase angle. We call this impedance the *capacitive reactance* X_c; so

$$Z_c = X_c = \frac{v}{i} = \frac{\text{A} \sin \omega t}{\text{A}\omega C \sin(\omega t + 90°)} = \left(\frac{1}{\omega C}\right)\angle -90° \; \Omega \qquad (1\text{--}42)$$

This impedance is a vector, having both a magnitude and a phase angle. With the voltage and current in phasor form, the equation is

$$X_c = \frac{V}{I} = \frac{\text{A}\angle 0°}{\text{A}\omega C\angle 90°} = \left(\frac{1}{\omega C}\right)\angle -90° \; \Omega \qquad (1\text{--}43)$$

This concept of impedance could be expanded to give impedances for other waveforms, but it has little practical value at this point in the text.

1.5.3 Inductors

Oersted discovered that when current flows in a conductor, magnetic flux is set up in and around the conductor. The strength of this field is proportional to the amplitude of the current. Oersted's law can be expressed as

$$\phi = ki \text{ W} \tag{1-44}$$

where ϕ is the flux in webers, k is a constant that depends on the construction of the coil, and i is the current in amperes. The flux is due to the motion of electrons and represents the energy involved in their flow. Because this energy is due to motion, it can be seen to resemble kinetic energy.

Faraday's law states that when magnetic flux cuts a conductor, a voltage is induced in the conductor. The voltage is proportional to the rate at which the flux cuts the conductor. Lenz's law states that the polarity of the induced voltage is always such as to oppose the current change that caused it.

An inductor is constructed by winding wire so that it is approximately helical. The coil may be formed on a ferromagnetic core, which provides an easy path for the magnetic flux and concentrates it within the coil. As a result, the changing flux set up by the current in any segment of the winding will cut most turns of the coil as it increases or decreases. Faraday's law gives the voltage of self-induction as

$$e = -N\frac{d\phi}{dt} \text{ V} \tag{1-45}$$

where N is the number of turns of the winding cut by the flux and $d\phi/dt$ is the rate at which the flux cuts the turns. The negative sign indicates that the induced voltage opposes the voltage that caused the current to flow in the first place. We usually consider the induced voltage to be a voltage drop. If so, its polarity is considered positive, and the voltage drop is given by

$$v = N\frac{d\phi}{dt} \text{ V} \tag{1-46}$$

The result is that a coil opposes any change in the current flowing through it. This effect is based on the energy stored in the field. Any variation in the current changes the flux. Since the energy is stored as a magnetic field, variation in the current changes the energy stored by the inductor. In Equation (1–12), energy is seen to be the time integral of power. Because a rapid change in energy requires a large amount of power, we can say that an inductor opposes changes in current flow.

The strength of opposition to current flow for any given coil is proportional to how effective the coil is in setting up flux. The latter depends on the size of the inductor and is called the *inductance*. Inductance is defined as

$$L = N\frac{d\phi}{di}\ \text{H} \tag{1-47}$$

with H denoting henrys. The derivative $d\phi/di$ is the rate of change of inductance with respect to current, in webers per ampere. An increase in the number of turns of a coil or in rate of change of flux with respect to current will cause an increase in inductance.

The voltage-current relationship in an inductance is readily found using some of the foregoing equations. The voltage drop in Equation (1–4b) can be multiplied by di/di to give

$$e_\text{L} = -N\frac{d\phi}{di}\cdot\frac{di}{dt}\ \text{V} \tag{1-48}$$

Solving Equation (1–47) for $d\phi/di$ and substituting the resulting expression into Equation (1–48) yields

$$e_\text{L} = -L\frac{di}{dt}\ \text{V} \tag{1-49}$$

This is the relationship between current and voltage for a pure inductance. The negative sign indicates that the current is entering the negative terminal. As discussed earlier, we usually consider the voltage to be a drop and ignore the negative sign, giving

$$v_\text{L} = L\frac{di}{dt}\ \text{V} \tag{1-50}$$

For a sine wave current input, such as

$$i = A\ \sin \omega t\ \text{A} \tag{1-51}$$

the voltage drop is

$$v_\text{L} = L\frac{d}{dt}(A\ \sin \omega t) = A\omega L\ \cos \omega t$$
$$= A\omega L\ \sin(\omega t + 90°)\ \text{V} \tag{1-52}$$

The impedance is

$$Z = X_\text{L} = \frac{e_\text{L}}{i} = \frac{A\omega L\ \sin(\omega t + 90°)}{A\ \sin \omega t} = \omega L \angle 90°\ \Omega \tag{1-53}$$

In phasor form,

$$I = A \angle 0° \text{ A} \tag{1-54}$$

and

$$V_L = A \omega L \angle 90° \text{ V} \tag{1-55}$$

So we can find the inductive reactance more easily by

$$X_L = \frac{V_L}{I} = \omega L \angle 90° \text{ } \Omega \tag{1-56}$$

Note that this impedance is a vector but not a phasor. Its angle is 180° different from that of a capacitive reactance. This means that the total reactance of a capacitor and an inductor connected in series is the difference of the two reactances.

We can find another expression for the voltage-current relationship in inductors by integrating both sides of Equation (1-50) and solving for the current. We obtain

$$\int_0^T v_L \, dt = Li + C_1$$

Solving for i and substituting $-C_0$ for C_1/L, we get

$$i = \frac{1}{L} \int_0^T v_L \, dt + C_0 \tag{1-57}$$

In this equation, C_0 is the current at the start of integration, $t = 0$ seconds. This can be called the initial condition of the inductor, since it represents the initial energy stored there.

Equations (1-50) and (1-57) give the relationship between current and voltage in inductors. We can say that they are the equivalent of Ohm's law for inductors.

1.5.4 Equivalent Circuits for Real Components

In a real circuit, resistance, capacitance, and inductance are never found as pure elements. A real element always has at least a small amount of each of the other types of impedance. In many if not most cases, the extraneous impedance effects are so small that they can be ignored.

At higher frequencies, there are cases where extraneous impedance effects must be considered. At higher frequencies, capacitive reactance is very small and stray capacitance can drastically change any impedance it is in parallel with. Also, at higher frequencies, inductive reactance is high. At such frequencies, the small inductance of the circuit wiring may be large enough to drop a significant amount of voltage.

These stray effects are, by their nature, usually distributed along or across the desired impedance. As a rule, though, they may be considered to be concentrated or lumped without significant error.

1.5.4.1 Resistance   The total impedance of a resistor includes reactive components. When current flows through a resistor, magnetic flux is set up, as it is with current flow in any conductor. This indicates that the resistor has a small inductance.

Also, the two resistor leads are conductors at different potentials. This potential difference causes an electric field to be set up between the leads. Therefore, there is a small capacitance between them.

These reactances are distributed throughout the resistor, but they can be approximated by lumped inductance and capacitance. The equivalent circuit is shown in Figure 1–5(a). The values of inductance and capacitance are very small, usually making their effects negligible below the very high-frequency (VHF) region.

1.5.4.2 Capacitance   The dielectric between the plates of a capacitor is an insulator, not a conductor, but the material does have a small number of free electrons. This means that it is possible for minute amounts of current to flow between the plates. There is also a conduction path across the edges of the dielectric sheet. The current that flows by these paths is called *leakage current*.

The capacitor has leads which, like the leads of resistors, have a small inductance as well as a small resistance. These effects are obviously distributed, but can again be considered to be lumped without appreciable error. They cause the equivalent circuit of a real capacitor to be as shown in Figure 1–5(b).

For most dielectric materials, the parallel resistance is very large and the series inductance and resistance are very small. The inductance has little effect at frequen-

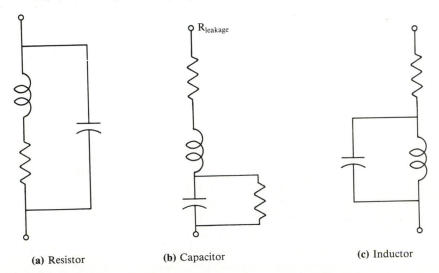

(a) Resistor   (b) Capacitor   (c) Inductor

FIGURE 1–5   Equivalent circuits of components

cies below the VHF range. For most dielectric materials, as long as the capacitance is not very large and the frequency is not too high, the effects of lead inductance and resistance and leakage current may be ignored.

1.5.4.3 Inductance A real inductor is not a pure inductance: it is made of a length of wire that has resistance. Also, there may be other power loss effects that contribute to the AC or effective resistance. The individual turns of the coil are conductors that have a dielectric insulation between them, so they add capacitance to the coil.

In a real inductor, the capacitance and resistance are distributed throughout the coil but can be considered to be lumped, as shown in Figure 1–5(c). The effect of the capacitance usually is negligible for inductors operating within their design frequency range. The resistance, however, is never so small that it can be ignored.

1.5.5 Admittance

Admittance is a way of expressing the current-voltage relationship in circuits that has been used extensively for parallel circuits. Admittance is the reciprocal of impedance; it is a measure of how easily current may be caused to flow in a circuit. Like impedance, admittance is a vector and is given by

$$Y = \frac{1}{Z} = \frac{1}{R + j(X_L - X_C)} \text{ S} \tag{1–58}$$

The units of admittance are siemens.

The admittance of an ideal resistor is called its *conductance*. The symbol used for conductance is G. Mathematically,

$$Y_R = \frac{1}{R \angle 0°} = G \angle 0° \text{ S} \tag{1–59}$$

The units of conductance are also siemens.

The admittance of a reactor is called its *susceptance*, B_L or B_C. Inductive susceptance and capacitive susceptance are the reciprocals of inductive reactance and capacitive reactance, respectively, and are given by

$$B_L = \frac{1}{X_L \angle 90°} = B_L \angle -90° \text{ S} \tag{1–60}$$

and

$$B_C = \frac{1}{X_C \angle -90°} = B_C \angle 90° \text{ S} \tag{1–61}$$

The units of susceptance are siemens.

The concept of admittance makes hand calculation of values in parallel circuits easier, but now that calculators are readily available, its usefulness has decreased. It does make certain equations easier to write, including those used in nodal analysis.

1.5.6 Mutual Inductance and Transformers

When the flux set up by an inductor cuts the turns of the inductor and induces a voltage in it, the phenomenon is called *self-inductance*. When the same flux cuts other conductors, a voltage will be induced in them, too, and this phenomenon is called *mutual inductance*. Inductors are often placed so that they can produce mutual inductance.

The symbol used for mutual inductance is *M,* and the units are henrys, the same units used to measure self-inductance. In any two coupled inductors, the magnitude of the mutual inductance is the same for either direction of power transfer.

The operation of transformers depends on mutual inductance. The amount of mutual inductance in a transformer is a function of how much of the flux from one coil cuts the other coil. When a large percentage of the flux produced by one coil cuts the other coil, we say that the coils are *closely coupled* or they have a *high coefficient of coupling*. The symbol used for the coefficient of coupling is *k*.

The relationship between the self-inductance of each coil, their mutual inductance, and the coupling coefficient is

$$M = k\sqrt{(L_1 L_2)}\ \text{H} \tag{1-62}$$

In this equation, L_1 and L_2 are the self-inductances of the coils and M is the mutual inductance, all in henrys. The coefficient of coupling, k, is a constant for a specific transformer and is equal to the fraction of the flux from the input coil that cuts the other coil. It is found by

$$k = \frac{\phi_m}{\phi_1} \tag{1-63}$$

where ϕ_1 is the flux set up by the current in the input, or primary, coil and ϕ_m is the amount of primary flux that cuts the output, or secondary, winding. The fluxes are given in webers.

The coefficient of coupling depends on the transformer design. If the windings are tightly coupled, k approaches unity; for more loosely coupled transformers, k will decrease, approaching zero as a lower limit.

There are two equivalent circuits that may be used in analyzing the operation of transformers. The proper one to use depends on the general characteristics of the transformer involved.

Figure 1-6 is a simplified drawing of what we can consider to be an ideal transformer. Such a transformer has no losses, and all of its flux remains in the ferromagnetic core, so that the coefficient of coupling is unity. This flux is ϕ_m, the mutual flux.

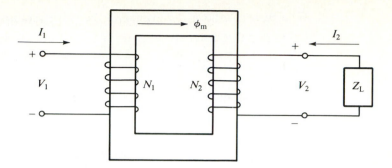

FIGURE 1–6   *Physical configuration of an iron core transformer*

The inductive reactance of both windings is assumed to be infinite. The input winding is called the primary, and the output side is called the secondary. We can use Equation (1–48) to find the voltage necessary to cause a specific rate of change in the amount of flux. For a transformer primary,

$$v_1 = N_1 \frac{d\phi_m}{dt} \ \text{V} \tag{1–64}$$

All of the flux in the core cuts the secondary winding, so

$$v_2 = N_2 \frac{d\phi_m}{dt} \ \text{V} \tag{1–65}$$

Dividing Equation (1–65) by Equation (1–64), we obtain the voltage ratio for ideal transformers:

$$\frac{v_2}{v_1} = \frac{N_2}{N_1} = a_v \tag{1–66}$$

From this equation, we see that the voltage step-up or step-down ratio of an ideal transformer is equal to the turns ratio of the transformer. Here, we have defined a_v as the voltage step-up ratio. In some books, a is defined as the current step-up ratio, the inverse of the ratio a_v.

If there are no losses in an ideal transformer, and if the impedances of the windings are very much larger than the load impedance, the input apparent power in volt-amperes must be approximately equal to the output apparent power. That is,

$$v_1 i_1 = v_2 i_2 \ \text{V-A} \tag{1–67}$$

Solving for the current ratio, we obtain

$$\frac{i_2}{i_1} = \frac{v_1}{v_2} \tag{1–68}$$

The current ratio is the inverse of the voltage ratio, or

$$\frac{i_2}{i_1} = \frac{N_1}{N_2} = \frac{1}{a_v} \tag{1–69}$$

In an ideal transformer, if the voltage is stepped up, the current is stepped down by the same ratio. The load impedance is then

$$Z_L = \frac{v_2}{i_2} \ \Omega \tag{1–70}$$

and the input impedance is

$$Z_{in} = \frac{v_1}{i_1} \ \Omega \tag{1–71}$$

We can find the ratio of the load impedance to the input impedance thus:

$$\frac{Z_L}{Z_{in}} = \frac{v_2/i_2}{v_1/i_1}$$

$$= (v_2/v_1)(i_1/i_2) = \left(\frac{N_2}{N_1}\right)^2 = (a_v)^2 \tag{1–72}$$

So the impedance ratio is equal to the voltage step-up ratio squared or the secondary-to-primary turns ratio squared. Equation (1–72) can be used to calculate the effect of all impedances on the secondary side of the transformer on the input impedance. From this equation, it can be seen that another function of transformers is to transform impedances. Indeed, they are often used to match impedances.

Figure 1–7 is a schematic of a loosely coupled transformer. An air core is shown, but some loosely coupled transformers have ferromagnetic cores. The figure shows the way mutual inductance may be indicated in a schematic drawing. M is the mutual inductance in henrys. The arrows show which coils have the mutual inductance between them.

The dot convention for transformers is also shown. The way the current flows in relation to the dots and the windings determines the sign of mutual impedance terms used to calculate the voltages in the circuit.

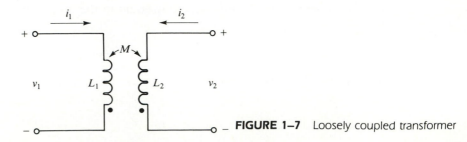

FIGURE 1–7   Loosely coupled transformer

If the current flows the same way in both windings, either from the dots to the windings or from the windings to the dots, the mutual terms are positive. If the current flows from a dot into one winding and from the other dot away from the other winding, the sign of the mutual term is negative. The dot convention is illustrated in Figure 1–8.

In both of the foregoing cases, the primary current enters the winding at the dot end and the secondary current leaves the winding at the dot end. As a result, the mutual inductance terms and the self-inductance terms have opposite signs.

Dots can be used to indicate the relative polarity of the terminals in cases when the load is largely resistive. If the phase shifts can be ignored, the voltages measured across each winding to the dotted terminals will be in phase.

An equivalent circuit that accounts for most of the linear characteristics of tightly coupled transformers is shown in Figure 1–9. In this circuit, R_p and R_s are the ohmic resistances of the primary and secondary windings, respectively, while L_p and L_s are the leakage inductances of those windings. Leakage inductance is that part of the total inductance which produces flux that does not cut the other coil. C_p and C_s are the capacitances between turns on the respective coils, C_w is the capacitance between the primary and secondary windings, R_c is a resistance that represents the

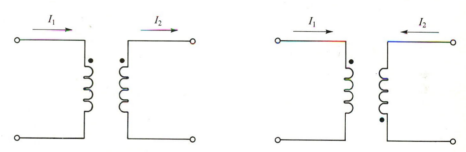

FIGURE 1–8 Use of the dot convention

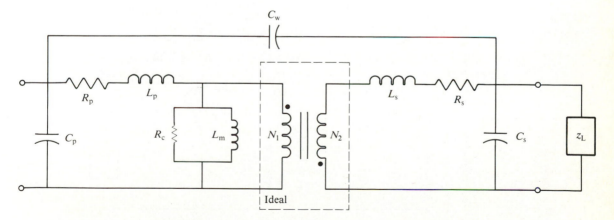

FIGURE 1–9 Practical equivalent circuit of iron core transformer

core losses, L_m is the magnetizing or mutual inductance of the transformer, and Z_L is the load impedance. Coupling between windings is simulated by an ideal transformer that has the turn ratio of the real transformer. In many cases, some of the equivalent circuit elements may be ignored.

If the transformer is a large power distribution transformer operating at or near full load, core losses are negligible. Tight coupling causes the leakage inductances to be minimal, also. The capacitive reactances at the operating frequency will be high in a well-designed transformer. They will have little effect on the operation of the transformer at that frequency. However, if transients occur, some of the frequencies involved may be much higher than the operating frequency. If so, the distributed capacitances may have such a low reactance at those frequencies that operation will be significantly affected.

Iron core transformers are inherently nonlinear devices, due to the effects of saturation. Under transient conditions, the transformer may very well be driven into saturation. When the transformer core is saturated, the peaks of the output wave will be flattened, indicating odd harmonic distortion. This distortion can produce significant amounts of the third harmonic in the output if the excitation is sinusoidal. Nonlinear effects are not analyzed in this text, but are mentioned to indicate that problems may occur.

Loosely coupled transformers may be represented by several simpler equivalent circuits. Figure 1–10 is one possible circuit diagram. When current flows in one winding, the mutual inductance produces a voltage in the other winding given by

$$v_{mx} = M \frac{di_y}{dt} \text{ V} \tag{1-73}$$

If we apply Kirchhoff's voltage law to the input circuit, we get voltage drops due to input current flow through the input resistance and inductance and due to the mutual term. The total drop is

$$v_1 = R_p i_1 + L_p \frac{di_1}{dt} + M \frac{di_2}{dt} \text{ V} \tag{1-74}$$

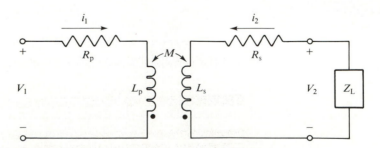

FIGURE 1–10 Practical equivalent circuit of loosely coupled transformer

In the output loop,

$$v_2 = R_s i_2 + L_s \frac{di_2}{dt} + M \frac{di_1}{dt} \text{ V} \tag{1-75}$$

The last term in each of these equations, $M \, di_2/dt$ and $M \, di_1/dt$, can be represented by a controlled source. Figure 1–11 gives the equivalent circuit, showing the drops across the coil resistances and self-inductances and the mutual terms. The signs of the mutual terms are positive because the assumed direction of current flow is from the dot through the coil in both windings.

EXAMPLE 1–3 Figure 1–12 shows a loosely coupled transformer. Draw the equivalent circuit using the transformer model previously described. Then apply Kirchhoff's voltage law to both the primary and secondary circuit.

We can first draw the equivalent circuit in the form illustrated in Figure 1–11, with the mutually coupled voltages as controlled sources. This is shown in Figure

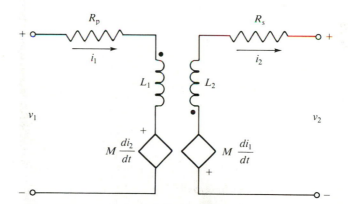

FIGURE 1–11   Equivalent circuit of loosely coupled transformer with controlled sources

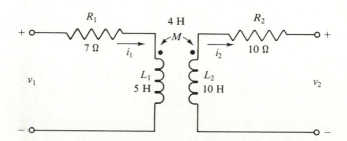

FIGURE 1–12   Loosely coupled transformer for Example 1–3

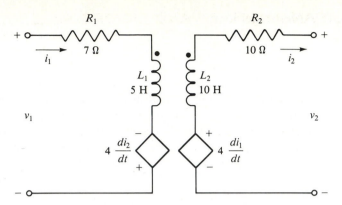

FIGURE 1–13 Equivalent circuit for Example 1–3

1–13. Then we write the two mesh equations, one for the input loop and one for the output loop. In the primary, the current goes from the dot into the winding. In the secondary, the current goes from the winding to the dot. Therefore, the mutual terms and the self-inductance drops will have opposite signs. The input equation is

$$v_1 = R_1 i_1 + L_1 \frac{di_1}{dt} - M \frac{di_2}{dt}$$

In the secondary, we have

$$v_2 = -R_2 i_2 - L_2 \frac{di_2}{dt} + M \frac{di_1}{dt}$$

Substituting the values from Figure 1–13, we obtain

$$v_1 = 7i_1 + 5 \frac{di_1}{dt} - 4 \frac{di_2}{dt}$$

and

$$v_2 = -10i_2 - 10 \frac{di_2}{dt} + 4 \frac{di_1}{dt}$$

Saturation of the core was mentioned as a possible source of harmonic distortion in iron core transformers. It also occurs in iron core inductors. Saturation is a nonlinear effect that reduces the rate of increase of flux as current in the coil increases. By Equation (1–47), this in turn will decrease inductance, possibly causing errors in calculating currents and voltages.

1.6   CIRCUIT ANALYSIS LAWS AND TECHNIQUES

In addition to Ohm's law, there are several basic laws and rules that are useful in circuit analysis. Among these are Kirchhoff's voltage and current laws, which have been described in Sections 1.3.2 and 1.3.3. The voltage divider and current divider rules simplify computations of individual variables in a circuit. Thévenin's and Norton's theorems simplify repetitive circuit analysis. Three methods can be used to analyze complex circuits: mesh and nodal analysis make use of Kirchhoff's laws in an organized manner and can be used on all linear circuits; superposition is used to simplify circuit analysis, particularly in the design of amplifier circuits.

1.6.1   Voltage Divider Rule

The voltage divider rule can be used to simplify the computation of voltages in a series string of impedances such as that illustrated in Figure 1–14. If the rule is used, the current need not be found first. Ohm's law and Kirchhoff's voltage law can be used to develop the equation for the voltage divider rule. In the figure, the current can be found to be

$$I = \frac{E}{Z_T} = \frac{E}{Z_1 + Z_2 + Z_3} \text{ A}$$

Using this current, we can find the drop across any one of the series impedances:

$$V_x = IZ_x \text{ V}$$

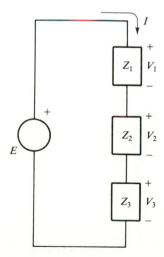

FIGURE 1–14   Voltage divider rule

Or, since $I = E/Z_T$ A,

$$V_x = \frac{EZ_x}{Z_T} \text{ V} \qquad (1\text{–}76)$$

This is the voltage divider rule in its general form. It agrees with Ohm's law in that, for a given current, the drops across impedances are proportional to the impedances.

EXAMPLE 1–4   In the circuit shown in Figure 1–15, find V_2, the voltage drop across R_2.
Summing the individual impedances, we find that the total impedance is 22 Ω. So

$$V_2 = \frac{EZ_2}{Z_T} = \frac{5(10)}{22} = 2.273 \text{ V}$$

To find V_2 without using the voltage divider rule, we find the current by dividing E, the voltage applied to the complete string, by the impedance of the string, R_T:

$$I = \frac{E}{R_1 + R_2 + R_3} = \frac{5}{5 + 10 + 7} = 227.3 \text{ mA}$$

Then Ohm's law can be used to find the drop across the individual resistor R_2:

$$V_2 = IR_T = (227.3 \times 10^{-3})(10) = 2.273 \text{ V}$$

For finding only one of the voltages in the series string, this method is a bit more complicated than using the voltage divider rule, since it involves two computations.

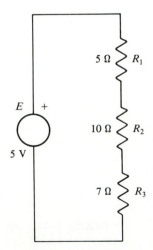

FIGURE 1–15   Circuit for Example 1–4

1.6.2 Current Divider Rule

The current divider rule is applied to impedances in parallel, to eliminate the need to solve for the voltage across the parallel elements when finding the currents through the branches. Two parallel impedances driven by an ideal current source are shown in Figure 1–16. The voltage across the parallel impedances is

$$V = I_1 Z_1 = I_2 Z_2 \; V$$

The impedance of the parallel pair is

$$Z_T = \frac{Z_1 Z_2}{Z_1 + Z_2} \; \Omega$$

So the voltage drop due to the source current flowing through the parallel pair is

$$V = \frac{I_T (Z_1 Z_2)}{Z_1 + Z_2} \; V$$

Thus,

$$I_1 Z_1 = I_2 Z_2 = \frac{I_T (Z_1 Z_2)}{Z_1 + Z_2} \; V$$

Solving for both I_1 and I_2, we get

$$I_1 = \frac{I_T Z_2}{Z_1 + Z_2} \; A$$

and

$$I_2 = \frac{I_T Z_1}{Z_1 + Z_2} \; A$$

More generally,

$$I_x = \frac{I_T Z_Y}{Z_x + Z_Y} \; A \qquad (1–77)$$

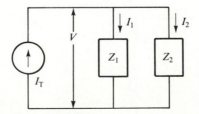

FIGURE 1–16 Current divider rule

where Z_Y is the total parallel impedance of all elements except that which the desired current I_x is flowing through. Since impedance is opposition to the flow of current, we should expect more of the total current to flow through any parallel element if the impedance in any of the other parallel components is increased. Equation (1–77) expresses this fact.

EXAMPLE   In Figure 1–17, find the current I_1 through resistor R_1.
1–5   Using the current divider rule, Equation (1–77), we have

$$I_1 = \frac{I_T Z_2}{Z_1 + Z_2} \text{ A}$$

Substituting the values from the figure,

$$I_1 = \frac{(10\angle 57°)(5\angle 90°)}{(10\angle 0° + 5\angle 90°)}$$

$$= 4.472\angle 120.4° \text{ A}$$

We can use a duality between two forms of Ohm's law to develop another form of the current divider rule. The AC form of Ohm's law is

$$V = IZ \text{ V} \qquad\qquad (1\text{--}78)$$

The voltage and current can be phasors; if so, the impedance must be expressed as a vector. Solving for current, we obtain

$$I = \frac{V}{Z} \text{ A}$$

The reciprocal of impedance is admittance. Substituting this into the foregoing equation, we get

$$I = VY \text{ A} \qquad\qquad (1\text{--}79)$$

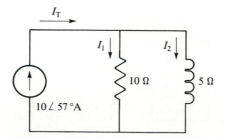

FIGURE 1–17   Circuit for Example 1–5

TABLE 1–1
Duality between Series and Parallel Circuits

Series	Parallel
V	I
I	V
Z	Y

This equation has the same form as Equation (1–78). They are considered duals in that a change in any of the related elements of either equation has the same effect as the corresponding change in the other equation. This duality is shown in Table 1–1. It works for any equation involving only the elements listed. For example, if we use it on Kirchhoff's voltage law, Equation (1–7), we replace voltage with current, getting

$$\Sigma I_n = 0 \text{ A}$$

which is Kirchhoff's current law, Equation (1–2).

The voltage divider rule is

$$V_x = \frac{E_T Z_x}{Z_T} \text{ V} \tag{1-80}$$

Using the duality table, we replace voltages by currents and impedances by admittances to get

$$I_x = \frac{I_T Y_x}{Y_T} \text{ A}$$

If we substitute impedances for the admittances, we get

$$I_x = \frac{I_T Z_T}{Z_x} \text{ A} \tag{1-81}$$

This is another form of the current divider rule.

EXAMPLE 1–6 Solve for the current I_1 in Figure 1–17 again, this time using the second form of the current divider rule. Then compare the result with that found by means of the first form.

Using the current divider rule, we get

$$I_1 = \frac{I_T Z_T}{Z_1} \text{ A}$$

where

$$Z_T = \frac{Z_1 Z_2}{Z_1 + Z_2} = \frac{(10\angle 0°)(5\angle 90°)}{(10\angle 0° + 5\angle 90°)}$$

$$= 4.472\angle 63.43° \text{ A}$$

or

$$I_1 = \frac{(10\angle 57°)(4.472\angle 63.43°)}{10\angle 0°} = 4.472\angle 120.4° \text{ A}$$

Both forms of the current divider rule give the same result. The voltage did not have to be computed in either solution, but the total impedance did have to be found in the second method. Both methods are readily expanded to handle circuits with more than two parallel impedances, by combining all the other impedances as an equivalent single impedance.

1.6.3 Source Conversion

Some methods of circuit analysis are simplified if the sources in the circuit are all voltage sources or all current sources. Any real voltage source has an equivalent real current source, and any real current source has an equivalent real voltage source. For the sources to be equivalent, they must provide the same voltage and current to any linear load.

Figure 1–18 shows a real voltage source and a real current source. If they are equivalent, they will cause equal currents to flow in an equal load impedance Z_L Ω. For the voltage source, we can find the current through the load using the general form of Ohm's law,

$$I_L = \frac{V}{Z_T} = \frac{V_S}{Z_V + Z_L} \text{ A}$$

Z_V is the series impedance of the voltage source, R_S in Figure 1–18(a).

The current in the load caused by the current source is found by applying the second form of the current divider rule, Equation (1–77):

$$I_L = \frac{I_S Z_I}{Z_I + Z_L} \text{ A}$$

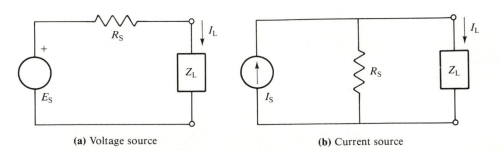

(a) Voltage source   (b) Current source

FIGURE 1–18   Real sources

Z_I is the parallel impedance of the current source, R_S in Figure 1–18(b).

Equating the expressions for these currents, we have

$$\frac{V_S}{Z_V + Z_L} = \frac{I_S Z_I}{Z_I + Z_L} \text{ A} \tag{1–82}$$

When the load is an open circuit, the voltages should be equal. For the voltage source, we have

$$V_{OC} = V_S \text{ V}$$

and, for the current source, we find that the open circuit voltage is

$$V_{OC} = I_S Z_I \text{ V}$$

Equating these, we obtain

$$V_S = I_S Z_I \text{ V} \tag{1–83}$$

From this equation, we can see that the numerators of Equation (1–82) are equal. So the denominators should be equal, too. Hence,

$$Z_V = Z_I \ \Omega \tag{1–84}$$

Consequently, equivalent real current sources and voltage sources have the same source impedances. We can usually assume that the reactances are negligible, so the source impedances are called R_S. Using this notation, we can rewrite Equation (1–84) as

$$R_S = Z_V = Z_I \ \Omega \tag{1–85}$$

Equation (1–83) then becomes

$$V_S = I_S R_S \text{ V} \tag{1–86}$$

Equations (1–85) and (1–86) can be used to find the equivalent of any simplified real source. Note that the polarity of the voltage source and the direction of the current from the current source must be such that current is caused to flow in the same direction in the load. Figure 1–19 shows equivalent current and voltage sources.

EXAMPLE 1–7   Convert the real current source in Figure 1–20(a) to the simplest possible real voltage source.

Using Equation (1–83), we find that the series impedance equals the original parallel impedance, $5K\angle75° \ \Omega$. The voltage is

$$E = IZ = (5 \times 10^{-3}\angle0°)(5 \times 10^{3}\angle75°) = 25\angle75° \text{ V}$$

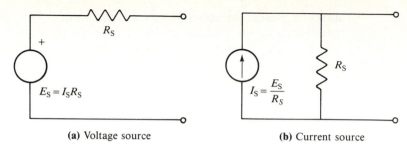

(a) Voltage source   **(b)** Current source

FIGURE 1–19 Equivalent voltage and current sources

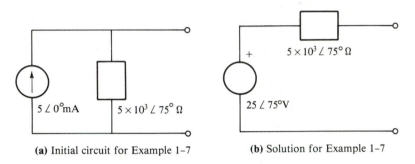

(a) Initial circuit for Example 1–7   **(b)** Solution for Example 1–7

FIGURE 1–20 Circuits for Example 1–7

Figure 1–20(b) shows the equivalent voltage source, consisting of an ideal voltage source in series with the impedance.

1.6.4 Mesh Analysis

Complex circuits, such as those having more than one source or those having elements that are neither in series nor in parallel, cannot be solved by simple series-parallel circuit analysis. Mesh analysis is one of several methods used to solve these circuits. Mesh analysis is also used in the analysis of transformer circuits.

In mesh analysis, we use Kirchhoff's voltage law to write a set of independent simultaneous equations. The unknowns in these equations are the currents in the circuit. The equations may be solved by hand using Cramer's rule, substitution, or Gaussian elimination. For systems involving four or more equations, computers and calculators with programs for the solution of simultaneous equations or with circuit analysis programs provide significant savings in time and effort.

Because each term in the simultaneous equations is a voltage term, it is easier to write the equations if all real current sources are converted to equivalent real voltage sources. A set of independent equations is easily set up if the circuit can be drawn in planar fashion. This implies that the circuit can be redrawn without crossovers.

Figure 1–21(a) shows a circuit with a crossover. This circuit can be redrawn in planar fashion as shown in Figure 1–21(b). Some circuits cannot be drawn in planar fashion, but these are practically nonexistent in the real world. An example of a nonplanar circuit is shown in Figure 1–22. With a circuit like this, it is difficult to know whether one has written a complete set of independent equations.

There is a simple test for the proper assignment of currents in the mesh equations for a circuit: the currents through elements that are not in series should not be identical. If they are, the set of equations is incomplete. Detailed methods for finding complete sets of independent equations can be found in a number of electrical engineering texts on network theory.

One method used to write mesh equations is called the *format method*. This method is systematic, easy to use, and less prone to error than other methods. The steps involved in using the method are as follows:

1.   Convert all real current sources to real voltage sources, for both dependent and independent sources.
2.   Draw the circuit in mesh form without crossovers.
3.   Assign a clockwise current to each mesh. There will be n such currents in a system having n meshes.

It may help to label the impedances with their voltage polarities, which should be positive on the side the assumed current enters.

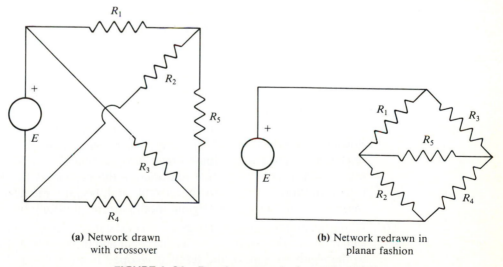

(a) Network drawn
with crossover

(b) Network redrawn in
planar fashion

FIGURE 1–21   Drawing networks for mesh analysis

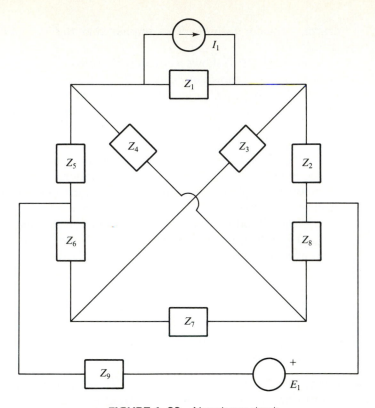

FIGURE 1–22 Nonplanar circuit

The preceding three steps are performed on the circuit as a whole. The following steps should be done on each mesh (for each mesh, one simultaneous equation will be developed):

4. Add all the impedances that any mesh current passes through, and then multiply by that mesh current. The resulting terms are positive. These impedances can be called the self-impedances.

5. In the same mesh, some of the impedances will have other mesh currents flowing through them. These impedances are sometimes called mutual impedances. For each of these other currents that flow in any branch of the mesh, sum up all the mutual impedances the current flows through in that mesh. Multiply each sum by the current, and, because the direction of flow is the reverse of that of the original mesh current, multiply by −1.

6. Going around the mesh in the direction of the mesh current, find the algebraic sum of the voltage sources, both dependent and independent. Equate this sum to the algebraic sum of the terms developed in steps 4 and 5. The resulting equation is one of the simultaneous mesh equations of the circuit. The general format will be

$$Z_{11}I_1 - Z_{12}I_2 - Z_{13}I_3 - \cdots - Z_{1n}I_n = \Sigma V_1 \text{ V}$$

$$-Z_{21}I_1 + Z_{22}I_2 - Z_{23}I_3 - \cdots - Z_{2n}I_n = \Sigma V_2 \text{ V}$$

$$\vdots \qquad \vdots \qquad \vdots \qquad \qquad \vdots \quad = \quad \vdots$$

$$-Z_{n1}I_1 - Z_{n2}I_2 - Z_{n3}I_3 - \cdots + Z_{nn}I_n = \Sigma V_n \text{ V} \qquad (1\text{--}87)$$

Here, I_1 through I_n are the unknown currents for meshes 1 through n, and Z_{11}, Z_{22}, through Z_{nn} are the sums of all the impedances in each mesh through which the individual mesh current flows. The other impedances, denoted by subscript digits that are not the same, are the mutual impedances. On the other side of the equals sign, the sum of the sources in mesh X is ΣV_X.

Mutual impedance Z_{XY} is the impedance common to both meshes X and Y and is equal to Z_{YX}. When the impedance values are entered into the mesh equations, the coefficients of the mutual terms are symmetrical about the diagonal formed by the self-impedance terms.

EXAMPLE 1–8   Write the mesh equations for the circuit shown in Figure 1–23.

Following the steps given on page 35, we can see that there are no current sources. The circuit is already in planar form. There are two meshes, so there will be two equations in two unknown currents. Currents I_1 and I_2 are assumed to be flowing clockwise as shown in the figure.

In the left-hand mesh, I_1 flows through the inductance and the resistor, so the I_1 term is $(R + jX_L)I_1$. In that mesh, I_2 flows through the inductance only, so the I_2 term is $-RI_2$. The forcing function is $+V_1$. In the right-hand mesh, the positive term is $(R - jX_C)I_2$ and the mutual term is $-RI_1$. The forcing function is $-V_2$. The two equations are, then,

$$(R + jX_L)I_1 - RI_2 = V_1 \text{ V}$$

$$-RI_1 + (R - jX_C)I_2 = -V_2 \text{ V}$$

These equations may be solved for the two currents by Cramer's method, substitution, or Gaussian elimination. Cramer's method is described in the appendix. In

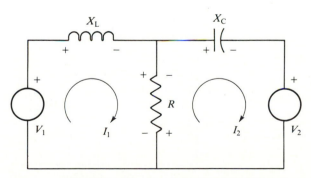

FIGURE 1–23 Circuit for Example 1–8

solving AC circuits, it is generally better to solve the equations for the unknowns before entering the numerical values of the phasors and vectors. Doing so reduces the chances of error in calculation.

If a circuit has a dependent or controlled source whose value depends on another current or voltage in the circuit, the magnitude must be expressed in terms of one of the unknown currents and combined with the other terms in that current.

If the dependent source is a current source, it should first be converted to a voltage source. Following the preceding procedure will place all sources on the right side of the equals sign. The controlled source terms must be expressed in terms of the unknown currents. Then the terms for the controlled sources must be combined with the other terms in the same currents. The array of current term coefficients may have lost their symmetry after this has been done.

EXAMPLE 1–9 The circuit shown in Figure 1–24 has a controlled source in its right-hand mesh whose magnitude is a function of the voltage across the resistor in the other mesh. Write the mesh equations for the circuit.

The first equation is

$$(R_1 + R_2)I_1 - R_2 I_2 = V_1 \text{ V}$$

The magnitude of the controlled source is $10\,V_X$. The second equation is

$$-R_2 I_1 + (R_2 - jX_C)I_2 = V_2 - 10\,V_X \text{ V}$$

V_X is the drop across R_1 and is equal to $R_1 I_1$. Substituting this in the equation, we get

$$10R_1 - R_2 I_1 + (R_2 - jX_C)I_2 = V_2$$

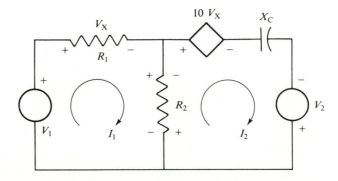

FIGURE 1–24 Circuit for Example 1–9

The two simultaneous equations can now readily be solved by Cramer's method. Notice that the original equations had symmetry in their mutual term coefficients, but this was destroyed when the final form of the second equation was developed.

EXAMPLE 1–10   It is necessary to find the output voltage of a loosely coupled transformer at a single frequency near the center of its frequency range. The input level can be assumed to be moderate, so that nonlinearity due to saturation can be ignored. The frequency is 1 radian per second. The circuit diagram is shown in Figure 1–25(a).

Using the development presented in Section 1.5.6, we have the equivalent circuit shown in Figure 1–25(b). We now can write the mesh equations. The mutual inductance terms have the same signs as the self-inductance terms due to the current directions with respect to the dots. Both currents flow from the dots through the windings.

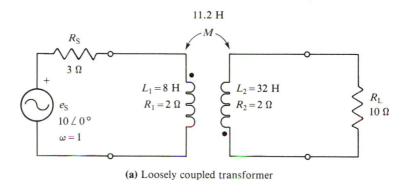

(a) Loosely coupled transformer

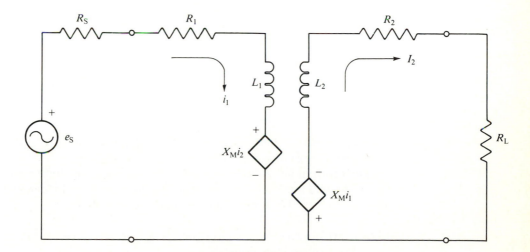

(b) Equivalent circuit

FIGURE 1–25   Circuit for Example 1–10

In the first mesh, we have

$$(R_S + R_1 + jX_{L1})I_1 = E_S - jX_M I_2 \text{ V}$$

The equation for the other mesh is even simpler:

$$(R_2 + R_L + jX_{L2})I_2 = -jX_M I_1 \text{ V}$$

Collecting terms with the unknown currents on the left-hand side of the equations, we get

$$(R_S + R_1 + jX_{L1})I_1 + jX_M I_2 = E_S \text{ V}$$

and

$$jX_M I_1 + (R_2 + R_L + jX_{L2})I_2 = 0 \text{ V}$$

Plainly, symmetry exists between the coefficients of the mutual terms of these equations.

The output voltage is $R_L I_2$, so we need to solve for I_2. Employing Cramer's method, we find that

$$I_2 = \frac{-jE_S X_M}{(R_S + R_1 + jX_{L1})(R_2 + R_L + jX_{L2}) - X_M^2} \text{ A}$$

Entering numerical values, we obtain

$$I_2 = \frac{-(10\angle 0°)(11.2\angle 90°)}{(3 + 2 + j8)(2 + 12 + j32) - 11.2^2} = -0.2709\angle 131.1° \text{ A}$$

The voltage across the load resistor is

$$V_L = I_2 R_L = (-0.2709\angle 131.1°)(12\angle 0°) = -3.2509\angle 131.1° \text{ V}$$

1.6.5   Nodal Analysis

Nodal analysis is useful for the same general type of circuit that mesh analysis is. In nodal analysis, a set of independent simultaneous equations is written by applying Kirchhoff's current law at junctions of the circuit. Here, the unknowns are the node voltages. As with mesh analysis, a format method is available that simplifies setting up the equations. The steps of this method are as follows:

1. Convert all real voltage sources into real current sources, for both independent and controlled sources.

2. Select a reference node. In an electronic circuit, this will usually be the ground terminal.

3. Label all the other nodes with their voltages. They will be numbered V_1 through V_{n-1} and there will be $n-1$ independent equations if there are n nodes.

The preceding three steps are performed on the circuit as a whole. The following steps must be done for each node separately to set up the equation for that node:

4. Add all the admittances connected to any one node. Multiply each sum by the voltage assigned to that node. The products will be positive. These admittances can be called the self-admittances.

5. For each other node, sum up all the admittances connected between the node whose equation is being developed and the other node. These are called mutual admittances. Multiply each by the voltage of the other node and by -1.

6. Sum up the currents entering and leaving the node. Currents entering the node are positive and currents leaving are negative. Equate these to the sums of the terms developed in steps 4 and 5 to give the $n-1$ nodal equations. These will take the form

$$
\begin{aligned}
Y_{11} V_1 - Y_{12} V_2 - Y_{13} V_3 - \cdots - Y_{1\text{m}} V_\text{m} &= \Sigma I_1 \text{ A} \\
Y_{21} V_1 + Y_{22} V_2 - Y_{23} V_3 - \cdots - Y_{2\text{m}} V_\text{m} &= \Sigma I_2 \text{ A} \\
\vdots \qquad \vdots \qquad \vdots \qquad\qquad \vdots \ &= \ \vdots \\
-Y_{\text{m}1} V_1 - Y_{\text{m}1} V_2 - Y_{\text{m}3} V_3 - \cdots + Y_{\text{mm}} V_\text{m} &= \Sigma I_\text{m} \text{ A}
\end{aligned}
\tag{1–88}
$$

where $m = n - 1$; V_1 through V_m are the voltages measured from each node to the reference node; admittances with subscripts of the form XX are the self-admittances, and those with subscripts of the form XY $(Y \neq X)$ are the mutual admittances; and ΣI_X is the algebraic sum of the current sources at node X.

As with the mutual impedances in the mesh analysis procedure, $Y_{XY} = Y_{YX}$. When the equations are first written, the coefficients are symmetrical about the diagonal formed by the self-admittances Y_{XX}. The result is that a system having n nodes is represented by $n - 1$ equations in $n - 1$ variables. The solutions provide the node voltages, measured with respect to the reference node.

EXAMPLE 1–11   Write the nodal equations for the circuit shown in Figure 1–26. The nodes are identified in the figure.

The first equation is

$$(G_1 + G_3 - jB_\text{L})V_1 - G_3 V_2 = I_1 \text{ A}$$

and the second is

$$-G_3 V_1 + (G_2 + G_3 + jB_\text{C})V_2 = -I_2 \text{ A}$$

We have used the admittances of the circuit, rather than the reciprocals of the impedances, in writing the equations. These simultaneous equations can readily be solved for the unknown voltages by Cramer's method.

If one of the sources of a circuit is a dependent source controlled by a current in the circuit, it must be expressed as a function of the unknown voltage or voltages that drive that current. For a solution, it must be added to those voltages in the equation. Note that doing so may remove the symmetry from the set of equations.

EXAMPLE 1–12   The circuit in Figure 1–27 has a controlled current source causing a current to flow from the node labeled V_1 to the reference node. The value of this current is $5I_X$, where I_X flows from the reference node to the node labeled V_2. Write the nodal equations for the circuit.

Using Ohm's law, we see that I_X is equal to V_2/R_2, or $V_2 G_2$. The equation for the currents at the first node is

$$(G_1 + G_3 - jB_L)V_1 - G_3 V_2 = I_1 + 5G_2 V_2 \text{ A}$$

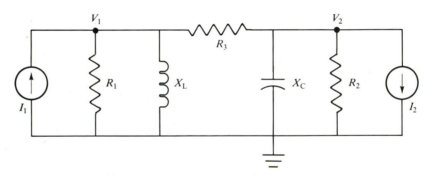

FIGURE 1–26   Circuit for Example 1–11

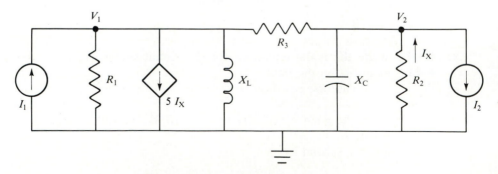

FIGURE 1–27   Circuit for Example 1–12

Collecting all terms in V_2 on the left side of the equals sign, we have

$$(G_1 + G_3 - jB_L)V_1 - (G_3 + 5G_2)V_2 = I_1 \text{ A}$$

At the second node, we have

$$-G_3 V_1 + (G_2 + G_3 + jB_C)V_2 = -I_2 \text{ A}$$

These two equations can be solved for the two unknowns V_1 and V_2. Again, symmetry existed in the original equations. It was lost when we shifted the controlled source term to the left of the equals sign. If the controlled source had been dependent on a current or voltage that occurred between the nodes labeled V_1 and V_2, symmetry would still exist.

1.6.6   Superposition

The superposition theorem states that the current through or voltage across a component in a circuit with several sources is equal to the sum of the currents or the voltages due to each individual source. To find the effect of an individual source, all other sources must be zeroed.

To zero an ideal voltage source, we replace it by an element which ensures that the voltage across its terminals is zero. That element is a short circuit. Similarly, to zero an ideal current source, we replace it by an element through which no current can flow—that is, an open circuit.

A procedure which will reduce errors that may be incurred in writing the equations is to redraw the circuit repeatedly with all but one source zeroed—a different one each time the circuit is redrawn. The result is a series of circuits, each having only one source. Each circuit can readily be solved for that part of the current or voltage desired using standard single-source circuit analysis techniques. The procedure is repeated for each source, and then the individual results are summed.

EXAMPLE 1–13

Find the current through the inductor in Figure 1–28 using superposition.

If we zero the current source, we have the circuit shown in Figure 1–29(a). We can find the current from the source if we first find the total impedance seen by the source. This is

$$Z'_T = R_1 + (jX_L) \| (R_2 - jX_C)$$

where the symbol $\|$ means "in parallel with." Substituting component values into the equation, we obtain

$$Z'_T = 10 + \frac{(j3)(5 - j7)}{j3 + 5 - j7} = 11.09 + j3.878 = 11.76 \angle 19.26° \ \Omega$$

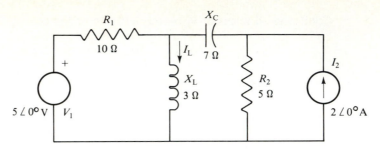

FIGURE 1–28 Circuit for Example 1–13

Using Ohm's law, we find the current from the source to be

$$I_T' = \frac{V_1}{Z_T'} = \frac{5\angle0°}{11.76\angle19.26°} = 425.2\angle-19.26° \text{ mA}$$

The current through the inductor caused by the voltage source can be found by using the current divider rule:

$$I_L' = I_T'\frac{R_2 - jX_C}{jX_L + R_2 - jX_C}$$

$$= (0.4252\angle-19.26°)\frac{(5 - j7)}{(j3 + 5 - j7)} = 571.2\angle35.06° \text{ mA}$$

To find the current component due to the current source, we need to zero the voltage source in Figure 1–28. The resulting circuit is shown in Figure 1–29(b). We then apply the current divider rule to find the current through the capacitor:

$$I_C'' = I_2\frac{R_2}{R_2 - jX_C + R_1\|jX_L}$$

$$= 2\angle0°\frac{5\angle0°}{5 - j7 + \dfrac{(10\angle0°)(3\angle90°)}{10 + j3}} = 1.387\angle36.10° \text{ A}$$

Next, we apply the current divider rule again, this time to find the current through the inductor:

$$I_L'' = I_C''\frac{R_1}{R_1 + jX_L}$$

$$= 1.387\angle36.10°\frac{10\angle0°}{10 + j3} = 1.329\angle19.40° \text{ A}$$

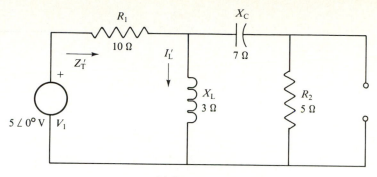

(a) Current source zeroed

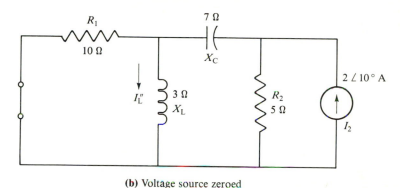

(b) Voltage source zeroed

FIGURE 1–29 Superposition circuits for Example 1–13

The total current is the sum:

$$I_L = I'_L + I''_L$$
$$= 0.4679 - j0.3276 + 1.254 + j0.4414$$
$$= 1.726\angle 3.781° \text{ A}$$

1.6.7 Choice of Method of Solution

In solving circuits by hand, it is highly desirable to choose the proper method of solution, not only to save time but also to reduce the chance of error. An important factor in this choice is which variable or variables must be solved for. With mesh analysis, we can find mesh currents directly, and this may be important, particularly when only one of the unknown currents flows through a specific component. Similarly, nodal analysis provides node voltages directly.

The structure of the network can be a factor. Circuits with a mesh structure, as illustrated in Figure 1–30(a), are usually more easily solved with mesh than with

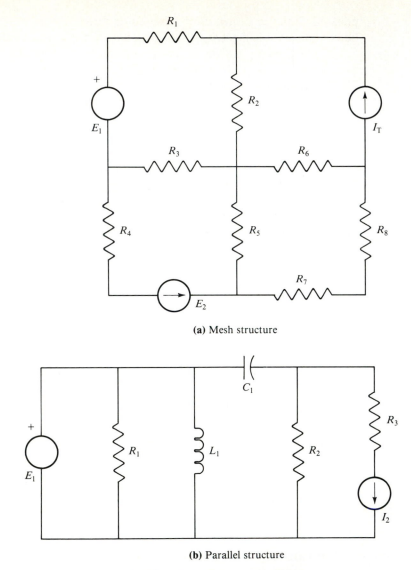

(a) Mesh structure

(b) Parallel structure

FIGURE 1–30   Circuit configurations

nodal analysis. If, on the other hand, the circuit has a generally parallel configuration, as shown in Figure 1–30(b), then nodal analysis will probably be simpler. As a rule, when a circuit has a parallel structure, fewer equations will need to be written with nodal analysis.

Superposition is often simpler when one voltage or current is needed in the solution. When a more complete solution is necessary, superposition is often complicated. Superposition is used in transistor design, where the DC or bias design must be completed first so that the transistor parameters can be determined.

The schematic diagram of a transistor circuit is shown in Figure 1–31. The DC design procedure makes it possible to find the parameters of the equivalent circuit. The coupling capacitors isolate the amplifier from the source and load impedances in the DC design. Some reasonable approximations make it possible to use the circuit shown in Figure 1–32 in the DC design.

One of the equivalent circuits for a transistor is shown in Figure 1–33. The values of the hybrid parameters are functions of the operating, or Q, point, which is determined in the DC design procedure. This equivalent circuit may be inserted in any circuit in place of the transistor.

The signal circuit design is performed on the resulting circuit, shown in Figure 1–34. In the design of AC amplifiers, the capacitors have a finite impedance, so the operation of the circuit is affected by the source and load impedances. For AC amplifiers, the signal and DC components are not added together after they are calculated. The DC component will be blocked by the coupling capacitor at the input to the next stage, so we can ignore it. In a DC amplifier, the DC output component due to the bias circuits cannot be removed without removing the signal, so it may appear in the following stage. Eventually, it must be compensated for, in any case.

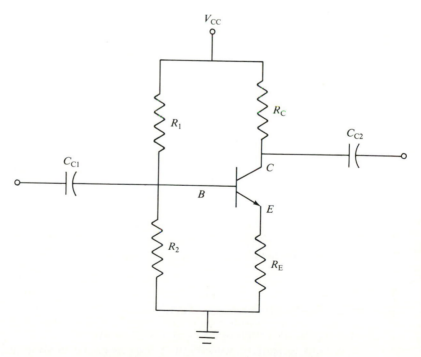

FIGURE 1–31 Transistor amplifier schematic

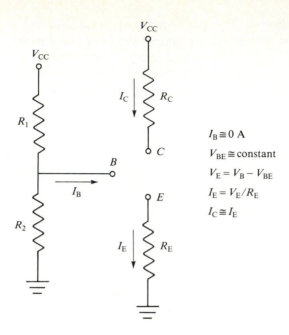

FIGURE 1–32 Simplified transistor bias circuit and approximations

$I_B \cong 0$ A

$V_{BE} \cong$ constant

$V_E = V_B - V_{BE}$

$I_E = V_E / R_E$

$I_C \cong I_E$

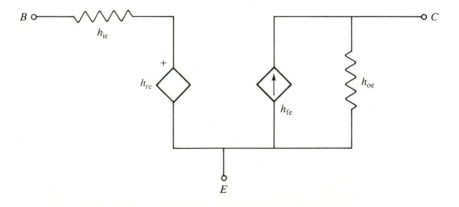

FIGURE 1–33 Hybrid equivalent circuit for transistor with common emitter connection

1.6.8 Thévenin's and Norton's Equivalent Circuits

In a circuit, it is sometimes desirable to analyze the effect as one component, such as a load resistor, is varied. In such a case, the analysis effort is reduced if the Thévenin or Norton equivalent for the part of the circuit external to the component to be varied is found first.

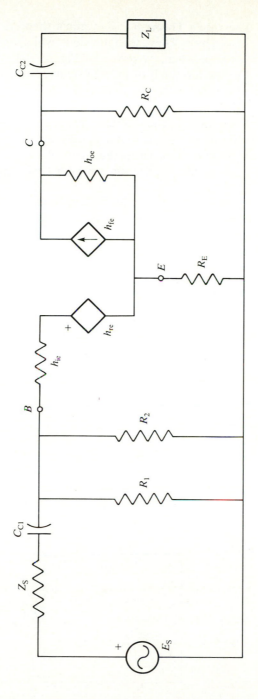

FIGURE 1–34 Transistor amplifier AC circuit

1.6.8.1 Thévenin's Theorem Thévenin's theorem states that any linear circuit may be replaced by its Thévenin equivalent, which consists of an ideal voltage source and a series impedance, as shown in Figure 1–35. If the external component is connected to the terminals of the Thévenin circuit, the circuit has the same effect on external devices as the original.

The Thévenin voltage is found by removing the component across which the equivalent is to be found and finding the open circuit voltage across the terminals. The Thévenin impedance is found by zeroing all sources in the circuit and finding the impedance appearing across the terminals. Zeroing the sources may be done by the methods described in Section 1.6.6 for each type of source.

EXAMPLE 1–14

Find the Thévenin equivalent for the circuit shown in Figure 1–36.

The Thévenin voltage V_{TH} is found by using the voltage divider rule to find the open circuit voltage at the terminals after the load impedance Z_L is removed. The circuit obtained from the rule is shown in Figure 1–37(a). The voltage is

$$V_{TH} = V_{oc} = V_1\frac{jX_L}{R + jX_L} = \frac{(5\angle0°)(10\angle90°)}{10 + j10} = 3.536\angle45° \text{ V}$$

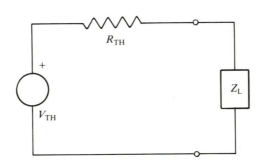

FIGURE 1–35 Thévenin equivalent circuit

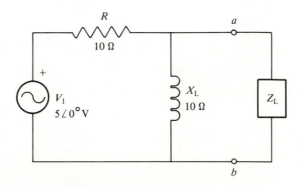

FIGURE 1–36 Circuit for Example 1–14

The circuit used to find the Thévenin impedance is shown in Figure 1–37(b). The Thévenin impedance consists of the inductance in parallel with the resistance, or

$$Z_{TH} = \frac{R(jX_L)}{R + jX_L} = \frac{(10\angle 0°)(10\angle 90°)}{10 + j10} = 7.071\angle 45° = 5 + j5 \ \Omega$$

The Thévenin equivalent is shown in Figure 1–38. The series impedance consists of a 5-ohm resistance in series with a 5-ohm inductance.

If a circuit includes a dependent source controlled by some variable within the circuit, the Thévenin impedance must be found in a different way from that given in Example 1–14. The impedance is then

$$Z_{TH} = \frac{V_{OC}}{I_{SC}} \ \Omega \tag{1–89}$$

the open circuit voltage divided by the short circuit current, both measured at the terminals across which the equivalent is to be found.

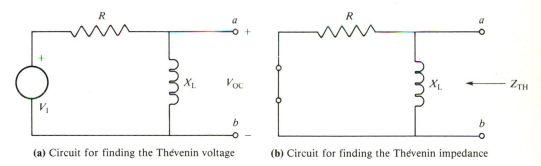

(a) Circuit for finding the Thévenin voltage   (b) Circuit for finding the Thévenin impedance

FIGURE 1–37   Circuits for Example 1–14

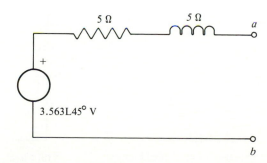

FIGURE 1–38   Thévenin equivalent circuit for Example 1–14

EXAMPLE 1–15 A circuit with a dependent source controlled by the voltage drop across a resistor in the circuit is shown in Figure 1–39. Find the Thévenin equivalent.

The open circuit voltage is found by calculating the current flow in the circuit shown in Figure 1–40(a). Applying Kirchhoff's voltage law, we have

$$V_1 - IR_S - jIX_L - 5IR_S = 10\angle 40° - 5I - j10I - 25I = 0 \text{ V}$$

Solving this for the current I, we obtain

$$I = \frac{10\angle 40°}{30 + j10} = 316.2\angle 21.57° \text{ mA}$$

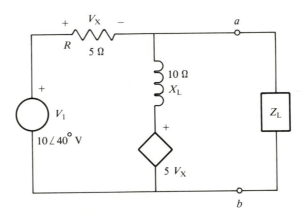

FIGURE 1–39 Circuit for Example 1–15

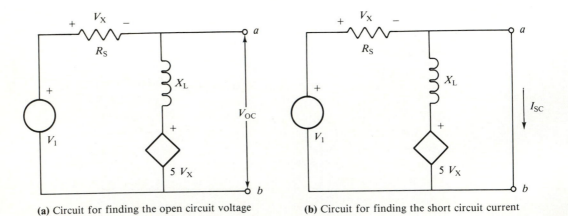

(a) Circuit for finding the open circuit voltage   **(b)** Circuit for finding the short circuit current

FIGURE 1–40 Circuits for Example 1–15

We find the open circuit voltage using Kirchhoff's voltage law again:

$$V_{OC} = V_{TH} = 5V_X + jIX_L = (25 + j10)(0.3162\angle 21.57°)$$

$$= 8.514\angle 43.57° \text{ V}$$

The short circuit current is found by calculating I_1 and I_2 for Figure 1–40(b) and then summing them. First, Ohm's law gives us

$$I_1 = \frac{V_1}{R_S} = \frac{(10\angle 40°)}{5} = 2\angle 40° \text{ A}$$

Then Kirchhoff's voltage law can be applied around the loop with the inductor, the dependent source, and the short circuit to give

$$5V_X - I_2 X_L = 0 \text{ V}$$

which can be solved for I_2:

$$I_2 = \frac{5V_X}{X_L} = \frac{5I_1 R_S}{X_L} = \frac{5(2\angle 40°)5}{10\angle 90°} = 5\angle -50° \text{ A}$$

Both I_1 and I_2 flow through the short circuit, so I_{SC} is their sum:

$$I_{SC} = I_1 + I_2 = 2\angle 40° + 5\angle -50° = 5.385\angle -28.20° \text{ V}$$

Finally, using Equation (1–89), we find that

$$Z_{TH} = \frac{V_{OC}}{I_{SC}} = \frac{(8.574\angle 43.37°)}{(5.385\angle -28.20°)}$$

$$= 1.592 \angle 71.57° = 0.5033 + j1.510 \text{ } \Omega$$

Thus, the Thévenin impedance is $0.5033 \text{ } \Omega$ resistive and $1.510 \text{ } \Omega$ inductive. The Thévenin equivalent circuit is shown in Figure 1–41.

1.6.8.2 Norton's Theorem As described in Section 1.6.3, the Norton equivalent is a real current source that is equivalent to the Thévenin circuit. The Norton equivalent is illustrated in Figure 1–42. The Norton current is equal to the current that would flow if the impedance across which the equivalent is to be found is replaced by a short circuit. In Section 1.6.3, it was found that the source impedance of the equivalent current source should be equal to that of the voltage source, so the Norton impedance is equal to the Thévenin impedance. That is,

$$Z_N = Z_{TH} = \frac{V_{OC}}{I_{SC}} \tag{1–90}$$

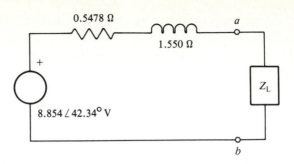

FIGURE 1–41 Thévenin equivalent circuit for Example 1–15

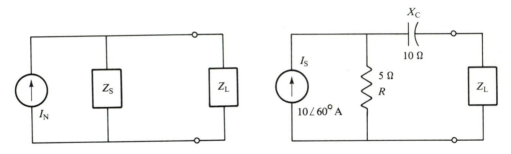

FIGURE 1–42 Norton equivalent circuit   **FIGURE 1–43** Circuit for Example 1–16

EXAMPLE 1–16   Find the Norton and Thévenin equivalent circuits for the circuit shown in Figure 1–43.

We can save effort by finding one of the equivalents directly and then converting to the other type of source. Recall that the savings are greater for more complex circuits. We can find the short circuit current by replacing the load impedance by a short circuit and calculating the current flowing in the short. The circuit is shown in Figure 1–44(a). The current is

$$I_N = \frac{I_S R}{R - jX_C} = \frac{(10\angle 60°)(5\angle 0°)}{5 - j10} = 4.472\angle 123.4° \text{ A}$$

The impedance can be found by using Equation (1–90), but it is simpler to zero the current source and find the impedance between the terminals, as shown in Figure 1–44(b). We have

$$Z_N = R - jX_C = 5 - j10 = 11.18\angle -63.43° \ \Omega$$

The Norton equivalent circuit is shown in Figure 1–45.
To find the Thévenin circuit, we find the voltage source that is equivalent to the current source. From Section 1.5.3,

$$V_S = I_S Z_I$$

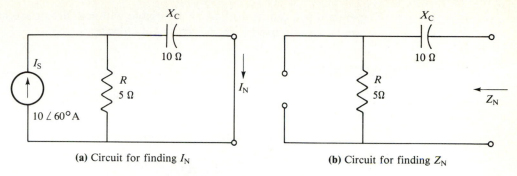

(a) Circuit for finding I_N

(b) Circuit for finding Z_N

FIGURE 1–44 Circuits for Example 1–16

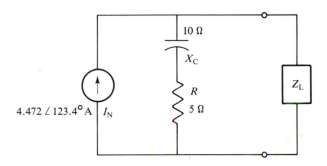

FIGURE 1–45 Norton equivalent circuit for Example 1–16

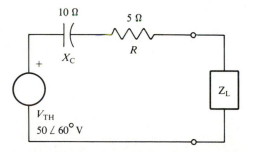

FIGURE 1–46 Thévenin equivalent circuit for Example 1–16

So

$$V_{TH} = (4.472\angle 123.4°)(11.18\angle -63.43°)$$

$$= 50\angle 60° \text{ V}$$

The impedance is the same as the Norton impedance. The Thévenin equivalent is shown in Figure 1–46.

Dependent sources controlled by a variable within a circuit are handled in a manner similar to that used with the Thévenin circuit.

PROBLEMS

1–1. Find i_r, i_e, and v_i for the circuit shown in Figure 1P–1.

1–2. Find v_r, v_i, and i_e in the circuit of Figure 1P–2.

1–3. Convert the following into phasor form:

$$i = 5 \cos(45t + 30°) \text{ A}$$

$$v = 25 \sin(32\pi t - 18°) \text{ V}$$

1–4. Convert the following phasor voltages to the time domain form:

$$V = 35\angle 17° \text{ V}, f = 35 \text{ Hz}$$

$$I = 18.5\angle -80° \text{ A}, f = 60 \text{ Hz}$$

1–5. Using Equation (1–38), find the current flowing in a 50-μF capacitor if $v_C = 4 \sin(45\pi t + 18°)$ V. Give your results in both time domain and phasor form.

1–6. Find the voltage across a 50-μF capacitor using Equation (1–37) if $i_C = 75 \sin 60\pi t$ A. Give your results in both phasor and time domain form.

1–7. Using Equation (1–50), find the voltage across a 1-mH pure inductor if $i_L = 75 \sin(350\pi t - 10°)$ A.

1–8. Using Equation (1–57), find the current that flows through a 1-H inductor in response to a voltage $v_L = 18 \sin(120\pi t + 35°)$ V.

1–9. A real resistor has a capacitance of 0.1 pF, an inductance of 0.3 μH, and a resistance of 100 Ω. What are the values of its total impedance at 75 MHz and at 75 GHz?

1–10. A real inductor has an inductance of 1 H, an effective AC resistance of 1×10^{-3} Ω, and a capacitance of 0.001 μF. What are the values of its total impedance at frequencies of 100 Hz and 10 MHz?

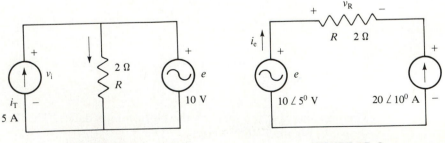

FIGURE 1P–1   FIGURE 1P–2

1–11. Figure 1P–3 shows a loosely coupled transformer circuit. The parameters of the circuit are as follows:

$$k = 0.8, \quad L_1 = 10 \text{ H}, \quad R_1 = 7 \text{ }\Omega, \quad L_2 = 5 \text{ H}, \quad R_2 = 3.5 \text{ }\Omega,$$
$$R_s = 0.5 \text{ }\Omega, \quad Z_L = 3 + j4 \text{ }\Omega$$

Redraw the circuit showing the dependent sources due to mutual inductance. Using Kirchhoff's voltage law, write one equation for each loop.

1–12. Figure 1P–4 shows a loosely coupled transformer with three windings. Redraw the circuit to show the controlled sources due to mutual inductance. Using Kirchhoff's voltage law, write an equation for each loop.

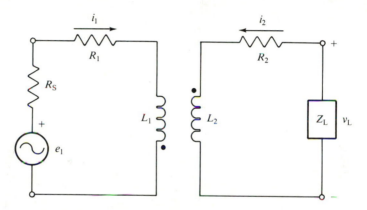

FIGURE 1P–3

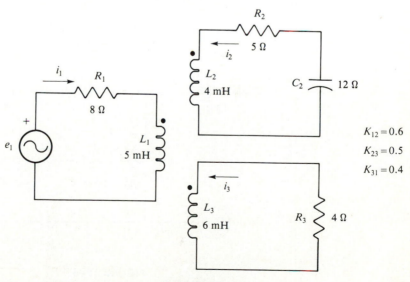

FIGURE 1P–4

1–13. Write the mesh equations for the circuit shown in Figure 1P–5. Solve them for the currents I_1 and I_2.

1–14. Write the mesh equations for each of the independent loops in Figure 1P–6. Solve for all currents.

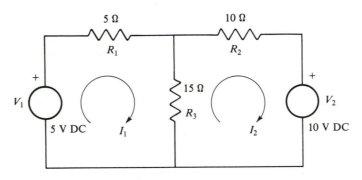

FIGURE 1P–5

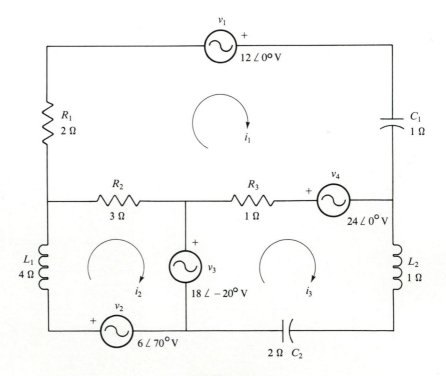

FIGURE 1P–6

1–15. Perform a nodal analysis on the circuit shown in Figure 1P–7. Find the node voltages.

1–16. Write the nodal equations for the circuit in Figure 1P–8. Solve for the node voltages.

1–17. Using superposition, find I_2 in the circuit shown in Figure 1P–9.

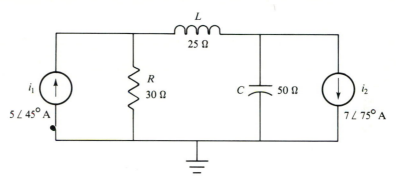

FIGURE 1P–7

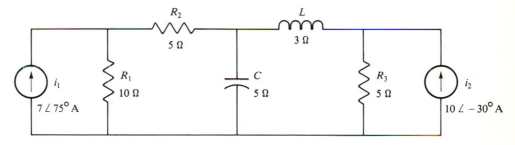

FIGURE 1P–8

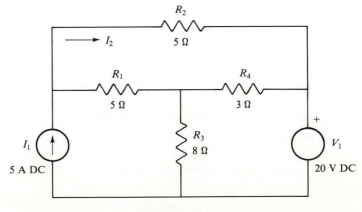

FIGURE 1P–9

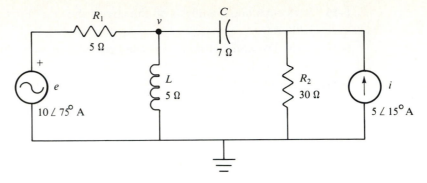

FIGURE 1P–10

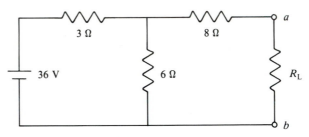

FIGURE 1P–11

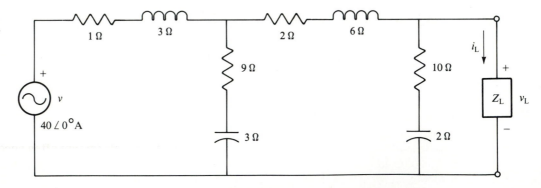

FIGURE 1P–12

1–18. Find V in the circuit shown in Figure 1P–10 using superposition.

1–19. Find V_{OC} and I_{SC} across terminals a and b in Figure 1P–11. Use these values to specify the Thévenin and Norton equivalents for the circuit. Use one of the equivalents to find the voltage across and the current through R_L when R_L equals 2, 4, and 6 Ω.

1–20. Find the Thévenin and Norton equivalents for the circuit across Z_L in Figure 1P–12. Solve for V_L and I_L if $Z_L = 5\angle 35°$ Ω.

CHAPTER 2

Waveforms

2.1 OBJECTIVES

On completion of this chapter, you should be able to:

- Express both repetitive and nonrepetitive waveforms in the form of equations. In many cases, the equations will be exact. In others, however, the equations will be an approximation to the actual waveform.

- Describe the relationship between many of the simpler and more common waveforms.

- Write equations for time-shifted waveforms and waveforms that are switched on at some specific time.

2.2 INTRODUCTION

In solving transient problems, a number of simple waveforms are often encountered. Some more complex waveforms are formed as a product or sum of two or more simple waveforms. On occasion, if an equation for a waveform is not known, it may be necessary to approximate it as the sum of several of the simpler functions. In order to find transient responses, it is desirable to know how to express waveforms by equations and how to perform mathematical operations on them.

2.3 SINE WAVES

The sine wave is one of the most important functions occurring in electrical systems. It is important for two reasons: first, AC power is generated as a sinusoidal function; second, all real repetitive and nonrepetitive functions can be represented by a sum of sinusoids.

From Chapter 1, a sine wave is represented by

$$y(t) = A_m \sin(\omega t + \theta)$$

where A_m is the peak magnitude, ω is the frequency in radians per second, and θ is the phase shift in degrees or radians. Since it is the product of radians per second and time, ωt is more easily expressed in radians. The phase angle θ is usually measured in degrees. The result is that the two angles are often expressed in different units. Nonetheless, in evaluating arguments of sinusoids, we must keep them in the same units.

The peak magnitude is usually given in volts or amperes. A sinusoidal function may be generated as shown in Figure 2–1. The vector shown in Figure 2–1(a) has a length of A_m and rotates at a constant angular velocity of ω radians per second with a phase angle θ at 0 seconds. The continuous sinusoid thereby generated and shown in Figure 2–1(b) will be equal to the vertical displacement of the vector as a function of time.

The period of the sine wave, T, is defined as one cycle of the function. This is the shortest time between any two points at which the waveform has the same amplitude and slope. The sine wave crosses the horizontal axis with a positive slope at the point where the argument of the sine wave, $\omega t + \theta$, equals $\pm 2n\pi$ radians, where n is zero or a positive integer. The rate of repetition of the sine wave is called its *frequency f* and is given by

$$f = \frac{1}{T} \text{ Hz} \qquad (2\text{–}1)$$

The unit of frequency is the hertz. When the frequency is one hertz, the signal has a repetition rate of one cycle per second. Since there are 2π radians per cycle, the repetition rate in radians per second is

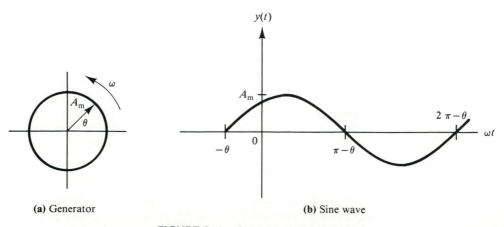

(a) Generator **(b)** Sine wave

FIGURE 2–1 *Generation of a sinusoid*

$$\omega = \frac{2\pi}{T} \text{ radians per sec} \tag{2-2}$$

This quantity is called the wave's *angular* or *radian frequency*.

The derivative of a sine wave is another sinusoid, the cosine. Figure 2–2 shows a sine wave with its derivative. Knowing that the derivative of a function is its slope allows us to use a simple technique to determine what the derivative of any sinusoid is: we can observe a sketch of the waveform and, after estimating its slope at several points, identify the specific sinusoid that is the derivative.

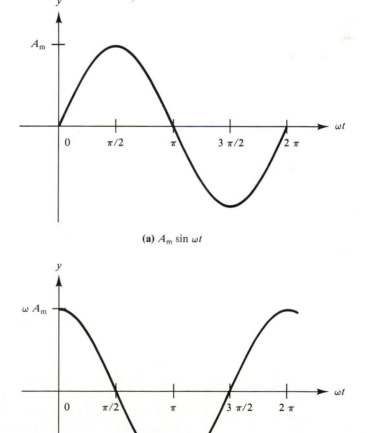

(a) $A_m \sin \omega t$

(b) $\omega A_m \cos \omega t$

FIGURE 2–2   Sine wave and its derivative

In Figure 2–2(a), the sine wave has zero slope at $\pi/2$ and $3\pi/2$ radians, so its derivative should be zero at those points. At 0, π, and 2π radians, the slope is steepest, being positive at 0 and 2π radians and negative at π radians. The cosine wave, shown in Figure 2–2(b), crosses the ωt-axis at $\pi/2$ and $3\pi/2$ radians and has peak positive values at 0 and 2π radians and a peak negative value at π radians, so we know that it is the only sinusoid that meets the requirements gleaned from an examination of the sine wave.

The magnitude of the derivative will not necessarily be the same as that of the original sine wave. If the frequency of the sine wave is high, its slope will be high. As a result, the amplitude of the derivative will be proportional to the sine wave frequency. The phase shift of the derivative should obviously be the same as that of the sine wave. Thus, if the sine wave is

$$y(t) = A_m \sin(\omega t + \theta) \tag{2–3}$$

then its time derivative is

$$\frac{dy}{dt} = A_m \omega \cos(\omega t + \theta) \tag{2–4}$$

Because an integral is the summation of contributions from the complete period of integration, it is more difficult to estimate directly than is the derivative. However, we do know that the integral is the inverse of the derivative. The continuous sine wave in Figure 2–2(a) must then be the integral of the cosine wave in Figure 2–2(b). If we shift both waves $\pi/2$ radians to the right and let $\omega A_m = B_m$, we get

$$y(t) = B_m \sin(\omega t + \theta) \tag{2–5}$$

as shown in Figure 2–3(a). The integral of this sinusoid is

$$\int y \, dt = -\frac{B_m}{\omega} \cos(\omega t + \theta) + C \tag{2–6}$$

as illustrated in Figure 2–3(b) if $\theta = 0°$. Again, if the sine wave has a phase shift, then its integral, the cosine wave, has the same phase shift. Note that in all of these waveforms the variable y can be a voltage or a current.

Both the derivative and the integral of a sine wave have a 90° phase shift with respect to the original wave. Differentiation causes a 90° phase lead with the amplitude multiplied by ω. Integration causes a 90° phase lag with the amplitude divided by ω.

EXAMPLE 2–1   Find the derivative of $y = 5 \sin(45t + 20°)$. The waveform is shown in Figure 2–4(a).
Using Equation (2–4), we find that the derivative of y is $dy/dt = 225 \cos(45t + 20°)$. This curve is shown in Figure 2–4(b). The derivative leads the original

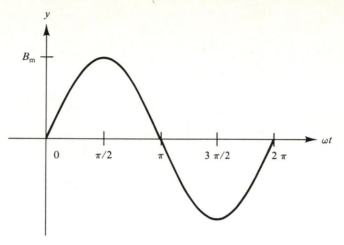

(a) $y = B_m \sin \omega t$

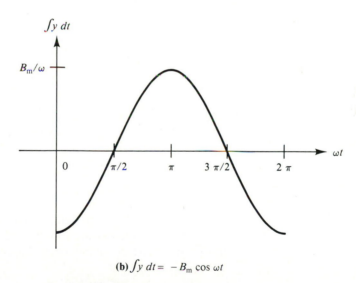

(b) $\int y \, dt = -B_m \cos \omega t$

FIGURE 2–3   *Sine wave and its integral*

wave by 90°, and its magnitude equals the original magnitude times the derivative of the argument.

EXAMPLE 2–2   Find the integral of $y = 75 \cos(5t - 25°)$. The waveform is shown in Figure 2–5(a). To use Equation (2–6), we need to convert the cosine to a sine wave. The sine wave lags the cosine by 90°, so we need to add 90° to the argument of the cosine function. This gives $75 \sin(5t + 65°)$. The integral is then $-15 \cos(5t + 65°) + C$. If C, the constant of integration, is zero, the result is as shown in Figure 2–5(b).

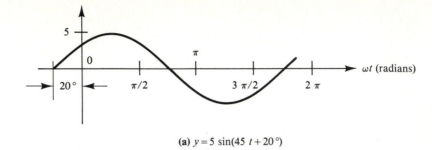

(a) $y = 5 \sin(45\ t + 20°)$

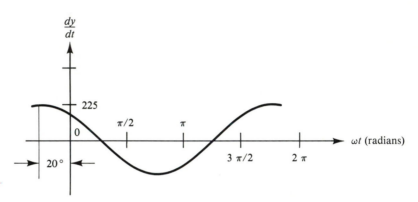

(b) $dt = 225 \cos(45\ t + 20°)$

FIGURE 2–4   Curves for Example 2–1

2.4 THE EXPONENTIAL FUNCTION

The exponential function is another function that often occurs in transient responses. The two forms of the exponential function are sometimes called the *growth* and *decay functions*. As often used in electrical engineering, the growth function is

$$y(t) = A\epsilon^{\alpha t} \tag{2–7}$$

The curve is illustrated in Figure 2–6. In electrical circuits, $y(t)$ is usually either voltage or current and A is the intercept of the function with the y-axis. The symbol ϵ denotes a fixed numerical value, 2.718281 . . . . In the exponent, α is called the *damping factor*. The growth function approaches zero as t goes to $-\infty$ and approaches $+\infty$ as time becomes infinite.

The exponential function is useful in scientific and engineering applications. It can be easily evaluated with scientific calculators. Passive electrical circuits cannot produce a growing exponential function, but the decay function is a common transient response.

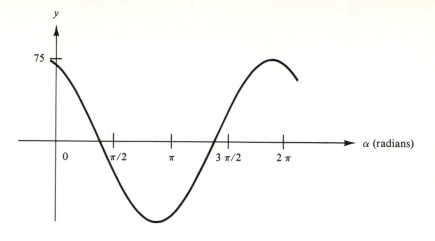

(a) $y = 75 \cos(5\ t - 25\,°)$

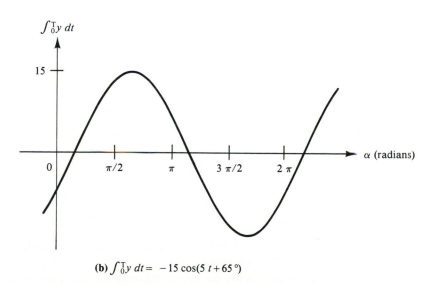

(b) $\int_0^T y\ dt = -15 \cos(5\ t + 65\,°)$

FIGURE 2–5 Curves for Example 2–2

The damping factor α is a positive constant that determines the amount that the amplitude of the growth function changes in a given length of time. In a time equal to $1/\alpha$, the amplitude will increase to ϵ, or approximately 2.718, times what it was at the beginning of the interval, starting at any point on the waveform. The exponential growth function increases without limit as time increases. It represents or is a factor in the output of an unstable system at least as long as it is operating in a linear part of its operating range. When α is large, the rate of change of either type of exponential factor is larger.

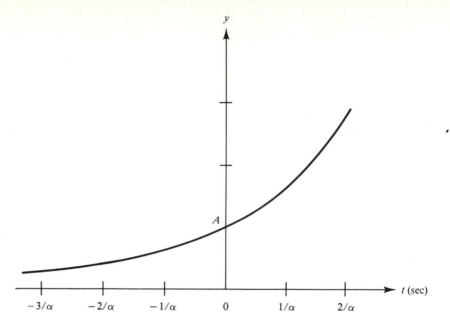

FIGURE 2–6 Exponential function, $y = A\epsilon^{\alpha t}$

If the exponent of the exponential function is negative, we have the decay function

$$y(t) = A\epsilon^{-\alpha t} \tag{2–8}$$

plotted in Figure 2–7. The function is infinite when t is $-\infty$, intercepts the y-axis at A, and approaches zero asymptotically as time approaches infinity. In $1/\alpha$ seconds, the amplitude of the function will decrease to $1/\epsilon$, or approximately 0.3679 times what it was at the beginning of the interval. A tangent drawn to the decaying exponential at any point intercepts the horizontal axis at $t = 1/\alpha$ seconds later, as shown in the figure. The decaying exponential function can be produced by passive circuits. It is one of the more common transient responses.

The derivative of the decaying exponential function is an exponential function with the same time constant:

$$\frac{dy}{dt} = \alpha A\epsilon^{-\alpha t} \tag{2–9}$$

When $t = 0$ seconds, the slope is $-A\alpha$. The integral of the decaying exponential function is

$$\int_0^t y \, dt = \frac{\epsilon^{-\alpha t}}{\alpha} + C \tag{2–10}$$

where C is the constant of integration.

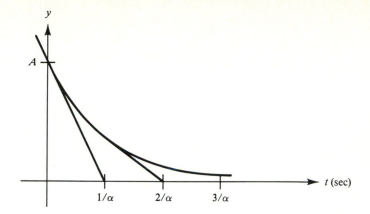

FIGURE 2–7   Plot of decaying exponential, $y = A\epsilon^{-\alpha t}$, showing tangents

The exponential function is used to describe scientific and financial processes in which the rate of change of a variable is proportional to its magnitude. Biological growth and decay and compound interest are among such processes. The decay process can be described by the simple differential equation

$$\frac{dy}{dx} = -By \qquad (2\text{–}11)$$

where B is a constant. As we can see from Equations (2–8) and (2–9), the decaying exponential function is a solution of this differential equation.

EXAMPLE 2–3   The decaying exponential $y = 5\epsilon^{-2t}$ that is switched on at the origin is shown in Figure 2–8(a). Find its derivative and its integral for $t > 0$ sec.

The time derivative of y, another exponential function, is found using Equation (2–9):

$$\frac{dy}{dt} = 5(-2)\epsilon^{-2t} = -10\epsilon^{-2t}$$

The integral is found from Equation (2–11):

$$\int_0^t y \, dt = \left(\frac{5}{2}\right)(1 - \epsilon^{-2t}) = 2.5(1 - \epsilon^{-2t})$$

The integral is thus a constant minus an exponential.

The derivative and integral waveforms are plotted in Figures 2–8(b) and 2–8(c).

The slope at any point on an exponential is directly proportional to the product of the amplitude at that point and the sign of the exponent. This means that the

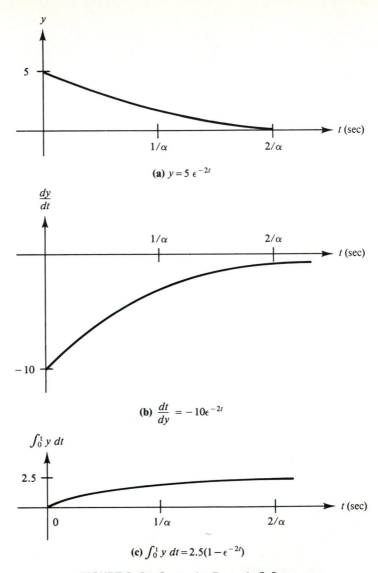

(a) $y = 5\,\epsilon^{-2t}$

(b) $\dfrac{dt}{dy} = -10\epsilon^{-2t}$

(c) $\int_0^t y\,dt = 2.5(1 - \epsilon^{-2t})$

FIGURE 2–8 Curves for Example 2–3

tangent to a decaying exponential at any point, say t_1, crosses the horizontal axis at $t_1 + 1/\alpha$. The amplitude of the exponential at the point where the tangent crosses the horizontal axis will be $1/\epsilon$ times the amplitude at the point of tangency. This is depicted in Figure 2–7.

EXAMPLE 2–4 Using your knowledge about the slopes and amplitudes of a decaying exponential function, sketch the exponential function $y = 10\epsilon^{-5t}$ for $t > 0$ sec.

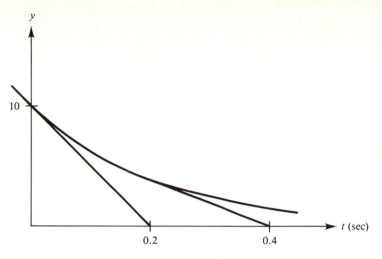

FIGURE 2–9 Curve for Example 2–4

The first thing to do is find the amplitude of the function when $t = 0$. This turns out to be $10\epsilon^0$, or 10. The curve intercepts the vertical axis at that value. Since $1/\alpha$ equals 0.2, at 0.2 second the amplitude is $10/\epsilon$, or 3.679. At that point, the curve is tangent to a straight line through the point and crossing the time axis 0.2 second later. Then, at $t = 2/\alpha = 0.4$ second, the amplitude is $10/\epsilon^2$, or 1.353. At this point, the curve is tangent to a line through the point and crossing the time axis at $t = 0.6$ second. The algorithm repeats itself at any point on the curve. Figure 2–9 illustrates the curve and two tangents, which start at points $1/\alpha$ seconds apart.

It may occur to the reader that no real system can produce an output that becomes infinite, as the exponential growth function does. If the output of a system began to increase without limit, the linear range of operation would be exceeded or perhaps components would burn out. In reality, in the transient analysis of passive or stable active systems, the decaying transient is the only one that occurs.

2.5 EULER'S EQUATION

A fundamental relationship exists between sinusoids and exponential functions that will become evident when we examine second-order linear differential equations with constant coefficients. The solutions of these equations can be either sinusoidal or exponential, depending on the relative values of the coefficients of the terms of the equations. The relationship between exponential and sinusoidal functions is expressed in Euler's equation:

$$\epsilon^{j\theta} = \cos \theta + j \sin \theta \tag{2–12}$$

From this equation, we can develop expressions for $\sin \theta$ and $\cos \theta$ in terms of exponential functions. Substituting $-\theta$ for θ in Equation (2–12), we obtain

$$\epsilon^{-j\theta} = \cos \theta - j \sin \theta \qquad (2\text{–}13)$$

Adding Equations (2–12) and (2–13) and solving for $\cos \theta$ yields

$$\cos \theta = \frac{\epsilon^{j\theta} + \epsilon^{-j\theta}}{2} \qquad (2\text{–}14)$$

If we subtract Equation (2–13) from Equation (2–12), we can solve for $\sin \theta$:

$$\sin \theta = \frac{\epsilon^{j\theta} - \epsilon^{-j\theta}}{2j} \qquad (2\text{–}15)$$

The relationship between exponential and sinusoidal functions involves imaginary numbers, so it is obviously not simple. We can use Euler's equation to develop the concept of phasors, which were briefly introduced in Chapter 1.

A vector or phasor may be described in two ways, as can be seen from Figure 2–10. In Chapter 1, phasors were described in terms of polar coordinates:

$$E = A_m \angle \theta \ \text{V}$$

It is also possible to use rectangular or Cartesian coordinates. In that case, the horizontal displacement is

$$a = A_m \cos \theta \qquad (2\text{–}16)$$

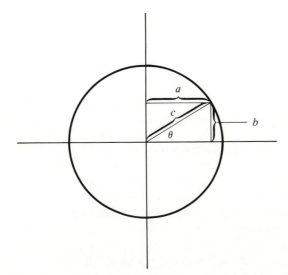

FIGURE 2–10   Phasor representation

while the vertical displacement is

$$b = A_m \sin \theta \tag{2–17}$$

The complete phasor is the vector sum of these:

$$E = a + jb = A_m \cos \theta + jA_m \sin \theta \tag{2–18}$$

Conversion between polar and rectangular coordinates can be done easily with a scientific calculator.

We can factor A_m out of the right side of Equation (2–18) and use Euler's relation, Equation (2–12), to express phasors compactly as

$$E = A_m \epsilon^{j\theta} \tag{2–19}$$

Here, A_m is the peak value of the wave and $\epsilon^{j\theta}$ represents the phase angle. This expression is equivalent to Equation (1–25) in Chapter 1.

As with other phasor expressions, we use a stationary phasor with an angle equal to the phase angle of the sinusoid at $t = 0$ seconds to represent the true phasor rotating at an angular rate of ωt radians per second. The true phasor is $A_m \epsilon^{j(\omega t + \theta)}$, of course. The actual voltage is the projection of this rotating vector onto the vertical axis.

2.6   THE EXPONENTIALLY DAMPED SINUSOID

The exponentially damped sinusoid is a common transient output waveform in circuits having two types of energy storage components. It is formed by replacing the amplitude factor of a sinusoid by an exponential function:

$$E = A \epsilon^{\alpha t} \sin(\omega t + \theta) \tag{2–20}$$

Because the exponential factor for passive circuits is negative, the result is a sinusoid with an exponentially decaying envelope. The exponentially damped sinusoid is illustrated in Figure 2–11.

Using Euler's equation, we can express the decaying exponential function in a way that emphasizes its complex exponential character:

$$E = A \epsilon^{\alpha t} \frac{\left[\epsilon^{j(\omega t + \theta)} - \epsilon^{-j(\omega t + \theta)} \right]}{2j} \tag{2–21}$$

In circuits with feedback, such as oscillators and automatic control systems, α can be positive or negative. When the exponent is positive, the envelope is a growing exponential, at least until the amplitude is limited by saturation.

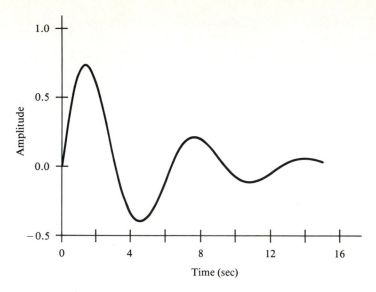

FIGURE 2–11 Exponentially decaying sinusoid

2.7 SPECIAL WAVEFORMS

There are three special waveforms that are useful in transient circuit analysis: the step function, the ramp function, and the impulse function. The basic step, ramp, and impulse functions have zero amplitude for $t < 0$ seconds, but may be shifted in time using the method described in the following subsection. Any one of these functions may appear alone or with other functions, either as a sum or a product.

2.7.1 The Step Function

The step function, the most important of the special functions, may be expressed by a pair of equations:

$$f(t) = 0, \, t < 0 \text{ sec}$$
$$f(t) = 1, \, t > 0 \text{ sec}$$

(2–22)

More simply, it can be expressed as

$$f(t) = U(t)$$

(2–23)

where $U(t)$ is the unit step function. With amplitude A, the equation is

$$f(t) = AU(t)$$

(2–24)

The units of the function can be volts or amperes. The unit step function is also called the *Heaviside unit function.*

The unit step function is shown in Figure 2–12. Note that there is a discontinuity at the origin, where the amplitude jumps from 0 to 1. This particular waveshape would be produced if a DC power source were suddenly connected to a circuit through a switch. The unit step function can also be a switching function used to turn on sinusoidal AC and other waveshapes. For instance,

$$f(t) = 120(\sqrt{2})U(t)\sin(120\pi t + \theta) \text{ V} \qquad (2\text{–}25)$$

is what would appear at the load side of the power switch of any household appliance if the switch were turned on at $t = 0$ seconds.

The step function is involved in producing the discontinuities that are responsible for the transient effects in most applications of Laplace transforms. It is sometimes called a switching function for this reason. Used as a multiplier, the step function can switch another waveform on. Similarly, we can express a switch that is turned off at time zero by

$$f(t) = [1 - U(t)]f_1(t) \qquad (2\text{–}26)$$

Here, the term $[1 - U(t)]$ switches $f_1(t)$ off at 0 seconds. An example of how the step function is used to switch off a DC voltage is shown in Figure 2–13. The 10-V DC signal is turned off at 0 sec. The equation that expresses this phenomenon is

$$f(t) = 10[1 - U(t)] \text{ V} \qquad (2\text{–}27)$$

At times, a waveform must be turned on or off at some time other than 0 seconds. This means that the switching step function must be shifted along the time axis. If it must be shifted to t_1 seconds, t in the argument of the step function is replaced by $t - t_1$. The step then occurs when the argument of the step function,

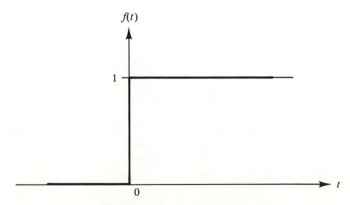

FIGURE 2–12 Unit step function

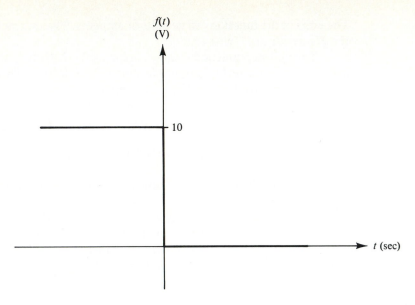

FIGURE 2–13 $f(t) = 10[1 - U(t)]$

$t - t_1$, is 0. If we want a waveform to start at 5 seconds, t_1 will be 5 seconds and every place the variable t occurs in the equation of the basic unshifted function, it must be replaced by $t - 5$.

The switching technique is useful for switching other functions on or off. It is commonly used with DC, sinusoids, and exponential functions. A square pulse can be produced by multiplying a DC level by the sum of a unit step and a delayed negative unit step. This turns on the DC level at $t = 0$ seconds and turns it off at the time of the delayed step. The pulse width is the time delay between the two steps.

EXAMPLE 2–5   Suppose a 10-volt, 2.5-second pulse, starting at $t = 5$ seconds, occurs as an input to a circuit. Express the pulse as a sum of step functions?

We can produce such a source by having a 10-volt source that is turned on at 5 seconds by $U(t - 5)$ and off by $U(t - 7.5)$. The equation is

$$E(t) = 10[U(t - 5) - U(t - 7.5)]$$

The pulse is illustrated in Figure 2–14.

EXAMPLE 2–6   How do we represent an unshifted 100-Hz sinusoid turned on for one cycle starting with the second alternation? Figure 2–15 shows the signal.

The signal is turned on after one-half of a period has occurred and is turned off one full period later. The period is

$$T = \frac{1}{f} = 10 \text{ msec}$$

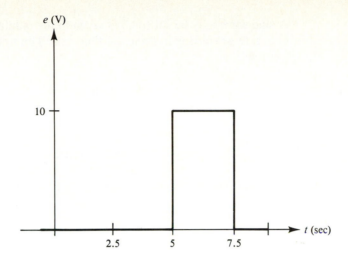

FIGURE 2–14   DC pulse for Example 2–5

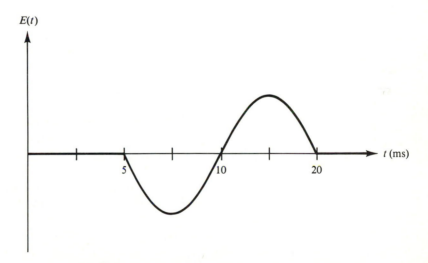

FIGURE 2–15   Switched sinusoid for Example 2–6

To represent the wave, we need to multiply the sinusoid, $\sin 100t$, by the sum of a positive unit step function delayed $T/2$, or 5, msec and a negative step delayed $3T/2$, or 10, msec. The following equation does the job:

$$E(t) = [U(t - 5 \times 10^{-3}) - U(t - 15 \times 10^{-3})] \sin 100t$$

Any waveform may be switched on or off in this way.

A sine wave may be shifted by two methods: adding an angle to the argument of its equation or shifting it along the time axis. The first of these is represented by the equation

$$y(t) = A_m \sin(\omega t + \theta) \tag{2-28}$$

and the second by the equation

$$y(t) = A_m \sin[\omega(t + t_1)] \tag{2-29}$$

In these equations, θ and ωt_1 refer to the same period of time.

EXAMPLE 2-7   Represent the waveform shown in Figure 2–16 using both the phase shift and the time shift method.

This waveform is a 10-volt-peak sine wave delayed by 20°. The period is 36 msec, so the wave is delayed by 2 msec. The frequency is the reciprocal of the period, 1/.036, or 27.78, Hz. Using the phase shift method of Equation (2–28), we get

$$e(t) = 10 \sin[2\pi(27.78)t - 20°]$$
$$= 10 \sin(174.5t - 20°)$$

The time shift method of Equation (2–29) gives us

$$e(t) = 10 \sin 2\pi(27.78)(t - 2 \times 10^{-3})$$
$$= 10 \sin(174.5)(t - 2 \times 10^{-3})$$

The first method is more commonly used.

The frequency content, or spectrum, of a positive step function is found to include all frequencies from DC on up. The spectral amplitude is inversely propor-

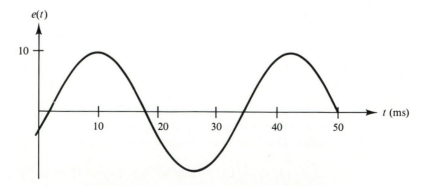

FIGURE 2–16   Shifted sine wave for Example 2–7

tional to the frequency. A common method of analyzing the operation of a high-fidelity loudspeaker is to measure its frequency response—that is, how much the sound level output of the speaker varies across its frequency band. Typical loudspeaker frequency responses have a number of closely spaced peaks and valleys. To approximate these variations adequately, many measurements must be taken across the audio band. The measurements are time consuming, but they are worth doing on the first few speakers produced. If, however, the loudspeaker is to be mass produced, their cost could be high if made on each speaker.

A comparatively simple test that could be done on each speaker is to apply a step function at its input and compare the response at the output with that produced with the original prototype loudspeaker. Any differences in shape of the output waves could alert the manufacturer to problems in quality control. This test would be much quicker and cheaper than performing a complete frequency response measurement on each speaker.

The technique has two problems, however. First, the spectral amplitude of the step function drops as frequency increases, so the measurement is not very sensitive to problems that cause the high-frequency response to change. The second problem is that a true step function is infinitely long and can only be approximated by producing a long pulse. We can provide a reasonably accurate test at low frequencies by using a pulse that is long enough so that its spectrum or frequency content has a significant magnitude at the lower edge of the band of frequencies we want the amplifier to reproduce. The high-frequency problem, on the other hand, can be solved only by using a different test function, the impulse function.

2.7.2 The Ramp Function

If we integrate the unit step function $U(t)$, we obtain what is called the *unit ramp function*:

$$\int_0^t U(t)\, dt = tU(t),\ t > 0$$
$$= 0, t < 0$$

(2–30)

The unit ramp function is zero for $t < 0$ seconds and has a slope of 1 unit per second for $t > 0$ seconds. To provide other slopes, we can use a constant multiplier. For example, the ramp function $AtU(t)$ has a slope of A units per second. Ramp functions can be used to represent or approximate different waveshapes.

EXAMPLE 2–8 The sweep voltage used in a cathode-ray oscilloscope is shown in Figure 2–17(a). It has a slope of 20 KV per second for 1 msec followed by a retrace slope of −200 KV per second for 100 μsec. Show how the voltage can be represented by a combination of ramp functions.

Figure 2–17(b) shows one way of representing the waveform. In this method, a number of individual ramp functions are summed up. The slope of the sweep

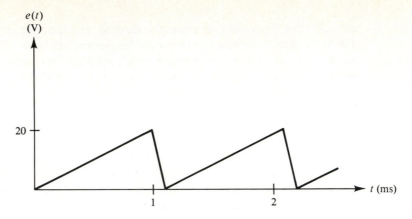

(a) Cathode-ray oscilloscope sweep voltage

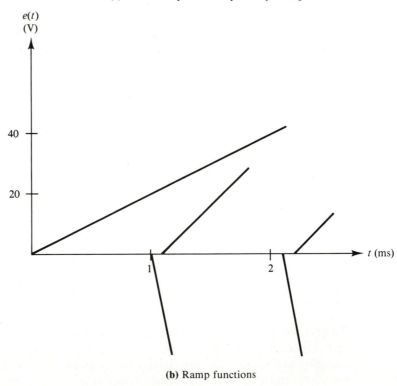

(b) Ramp functions

FIGURE 2–17 *Curves for Example 2–8*

voltage at any time will be the sum of the slopes of all the ramp functions summed together at that time.

From 0 to 1 msec, the slope of the sweep voltage is 20 KV per second. The first ramp function in the figure has the proper slope and magnitude for that segment. Its equation is $e = 20 \times 10^3\, tU(t)$.

From 1 to 1.1 msec, the sweep voltage has a slope of −200 KV per second. Since the original ramp function in the figure is continuing, we must add a shifted negative ramp function with a slope such that the sum of the two slopes is −200 KV per second. This new ramp will have a slope of −220 KV per second and a time shift of 1 msec. Its equation is

$$e = -220 \times 10^3(t - 10^{-3})U(t - 10^{-3})$$

Starting at 1.1 msec, the slope of the sweep voltage is again 20 KV per second. The sum of the slopes in the figure just before that time is −200 KV per second. So we must add a shifted ramp function whose slope is 220 KV per second to have the proper slope of 20 KV per second when $t \le 1.1$ ms. The unshifted ramp function is $e = 220 \times 10^3 \, tU(t)$. We can shift it to start at 1.1 ms by subtracting 1.1 ms from t each time it occurs in the equation for the waveform, giving $e = 220 \times 10^3$ $(t - 1.1 \times 10^{-3})U(t - 1.1 \times 10^{-3})$. The sequence from 2.1 msec on is alternately to add a negative slope ramp function after 1 msec and then, 0.1 msec later, add a ramp function with a positive slope. The final function is

$$e(t) = 20tU(t) - 220(t - 10^{-3})U(t - 10^{-3})$$
$$+ \ 220(t - 1.1 \times 10^{-3})U(t - 1.1 \times 10^{-3})$$
$$- \ 220(t - 2.1 \times 10^{-3})U(t - 2.1 \times 10^{-3})) + \cdots \text{ KV}$$

The terms after the first ramp function alternate between a positive and an equal negative slope.

2.7.3 The Impulse Function

The impulse function is very useful in transient analysis. It can be used to represent the initial state of the charge on capacitors or current in inductors. The concept is also important in the analysis of a circuit by Laplace transforms, as well as in circuit testing. By definition, the unit impulse function has zero width, infinite amplitude, and unit area. For nonunit impulse functions, the area is called the *strength* or the *magnitude*. The symbol used for a unit impulse function in the time domain is $\delta(t)$.

The integral of the unit impulse function is the unit step function. As integration is performed along the time axis, the integral is 0 until the impulse function occurs. At that point, since the impulse function has unit area and zero width, the integral jumps to a constant value of 1. We thus have

$$\int_0^t \delta(t)\, dt = U(t) \tag{2–31}$$

Figure 2–18 shows the unit impulse function and its integral, the unit step function. The arrowhead on the impulse function indicates that the amplitude is infinite. The strength of the impulse is indicated by the numeral 1 written beside the

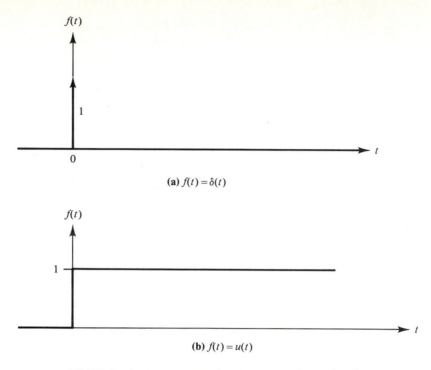

FIGURE 2–18 Unit impulse function and unit step function

impulse. Note that the three special waveforms are all related in that the derivative of the ramp function is the step function and the derivative of the step function is the impulse function. The latter can be seen by approximating the impulse function by a narrow pulse and then making it narrower while keeping its area constant. We can find its integral as the width approaches a limiting value of zero.

Figure 2–19(a) shows several steps in this process. All three pulses start at the same time and have unit area or strength, so as the pulse narrows its height increases. In the limit, when the pulse width becomes zero, the height is infinite. It is instructive to integrate these pulses. Figure 2–19(b) shows that the integral approaches a step function as the pulse width approaches zero.

Earlier, the statement was made that the impulse function was useful for testing. The frequency spectrum of the impulse function is very broad, being constant in amplitude at all frequencies, rather than decreasing in amplitude with increasing frequency as the spectrum of the step function does. So it would appear to be a better choice than the step function for a quick check on amplifier operation.

But it is impossible to produce a pulse with infinite amplitude and zero width. Fortunately, we can approximate the impulse function and test a specific circuit with a limited bandwidth by producing a pulse that is so narrow that the circuit cannot respond significantly while the pulse is on. Then the frequency content of the narrow pulse within the circuit bandwidth will be almost identical to that of the impulse

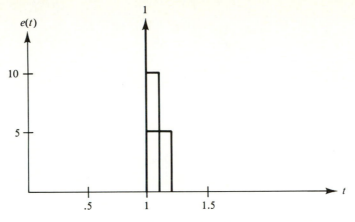

(a) Development of shifted unit impulse function $\delta(t-1)$

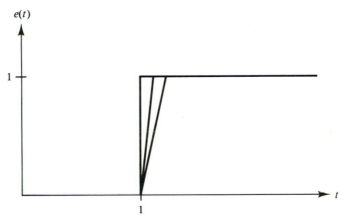

(b) Integrals of impulse function and approximations

FIGURE 2–19   Development of unit impulse

function, so the response of the circuit should be essentially the same. However, the response will be low in amplitude, which may prevent any effective use of the testing method.

2.7.4 Complex Waveforms

In practice, more complex waveforms than those discussed may be encountered. If the equation of the waveform is known, transient analysis can be readily performed. If the equation is unknown, but the actual shape of the waveform is known, it may be approximated by the use of a sum of ramp functions, as illustrated in Example 2–8 and Figure 2–17(b) in the modeling of a cathode ray tube sweep voltage by a group of ramp functions. For general functions, a reasonable approximation is

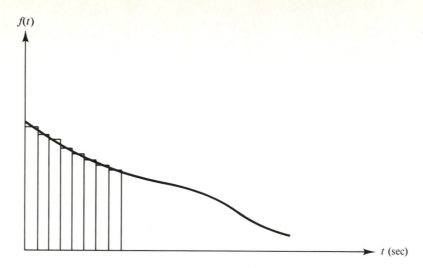

FIGURE 2–20 *Approximating a waveform by small blocks*

possible with this technique, as long as the sum of the ramps is kept close to the original function.

It may be simpler to approximate the waveform by a series of contiguous rectangular pulses, as shown in Figure 2–20. To give a reasonably accurate approximation, the pulses must be narrow enough so that the system will respond very little to an individual pulse while it is on. (Recall the earlier approximation of the impulse function.) Then the frequency components of the difference between the original signal and the approximation will be very much attenuated by an amplifier.

PROBLEMS

2–1. Sketch one cycle of the following sinusoids:
    a. $175 \sin 36\pi t$
    b. $72 \cos 40\pi t$

2–2. Sketch one cycle of the following sinusoids:
    a. $10 \sin(120\pi t + 45°)$
    b. $15 \sin(30\pi t - \pi)$

2–3. Write the expressions for the sinusoids shown in Figures 2P–1 and 2P–2. Give your answers as sine waves.

2–4. Write the expressions for the curves shown in Figures 2P–3 and 2P–4. Give your answers as sine waves.

2–5. Find the derivatives of the following sinusoids. Express the answers as sine waves.
    a. $v(t) = 120 \sin 120\pi t$ V
    b. $i(t) = 13 \cos(800\pi t - 35°)$ A

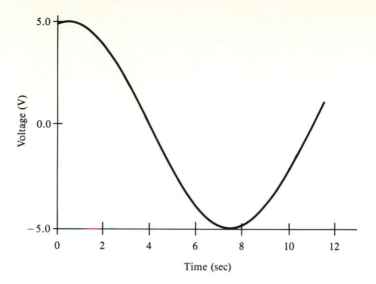

FIGURE 2P–1

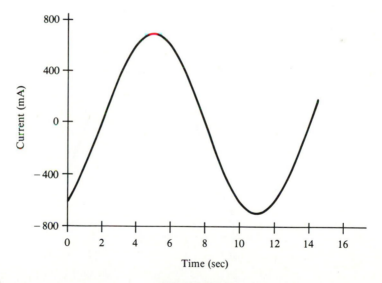

FIGURE 2P–2

2–6. Find the derivatives of the following in terms of sine waves:
 a. $v(t) = 1.745 \sin(36\pi t + 17°)$ V
 b. $i(t) = 207.8 \cos(100t - 18°)$ A

2–7. Integrate the following, giving the results as sine waves. Assume that the constant of integration is zero.

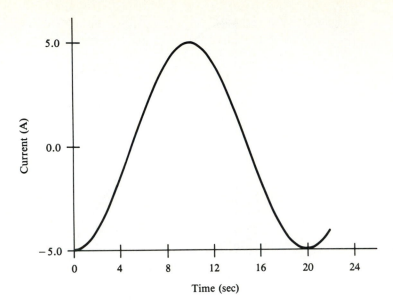

FIGURE 2P–3

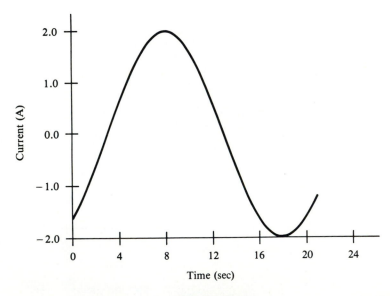

FIGURE 2P–4

   a. $v(t) = 18 \cos 90t$ V
   b. $v(t) = 190 \sin(35t + 49°)$ V

2–8. Integrate the following, giving the results as sine waves. The constant of integration is zero.
   a. $i(t) = 1000 \cos(39t - 25°)$ A
   b. $v(t) = 790 \sin 85t$ V

2–9. Sketch the following curve, using the relative amplitudes given in the text. Sketch the curve for two time constants after $t = 0$ sec.

$$e(t) = 20\epsilon^{-20t} \text{ V}$$

Show graphically that the tangent at the time t_1 seconds on the curve intercepts the time axis at $t = t_1 + \tau$ seconds.

2–10. Find:

a. $v(t) = \dfrac{d}{dt}(17\epsilon^{-5t}) \text{ V}$

b. $i(t) = \displaystyle\int_0^t 25\epsilon^{-7t}\,dt \text{ A if } i(0) = 0 \text{ A}$

2–11. Write expressions for the signals shown in Figures 2P–5, 2P–6, and 2P–7.

2–12. Sketch the following:

a. $v(t) = 32U(t - 5) - 64U(t - 10) + 96U(t - 15) - 32U(t - 20) \text{ V}$
b. $v(t) = 10U(t) \sin 35t - 20U(t - 4.574 \times 10^{-3}) \sin 35t \text{ V}$

2–13. Develop expressions for the functions shown in Figures 2P–8, 2P–9, and 2P–10.

2–14. Write the expressions for the derivatives of the functions in Problem 2–13, and sketch their waveforms.

2–15. Sketch

$$v(t) = 6t + 12(t - 6) - 10(t - 12)$$

over a range of time from 0 to 18 seconds.

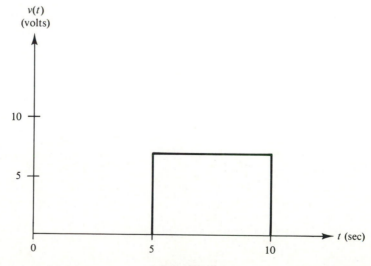

FIGURE 2P–5

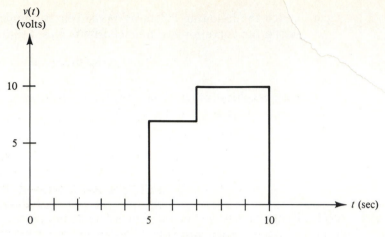

FIGURE 2P–6

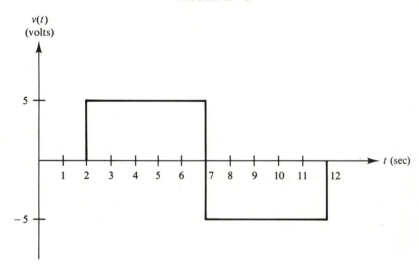

FIGURE 2P–7

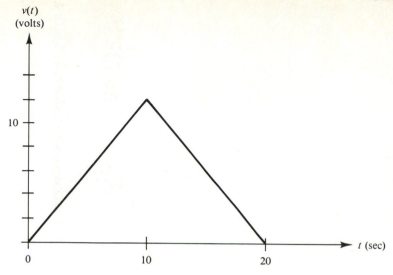

FIGURE 2P–8

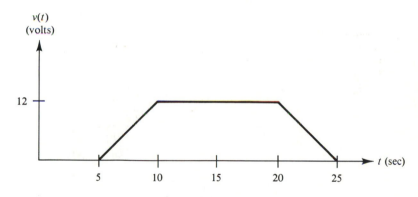

FIGURE 2P–9

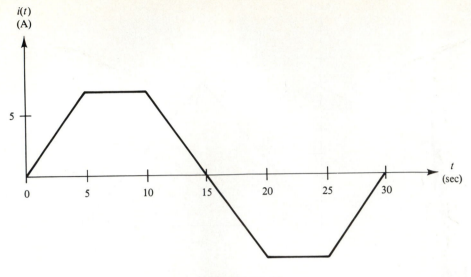

FIGURE 2P–10

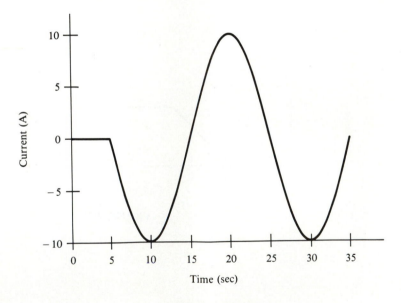

FIGURE 2P–11

2–16. Express the waveform shown in Figure 2P–11 in two ways. The curve is a sinusoid.

2–17. Write the expressions for the waveforms shown in Figures 2P–12 and 2P–13.

2–18. Write equations for the functions shown in Figures 2P–14 and 2P–15.

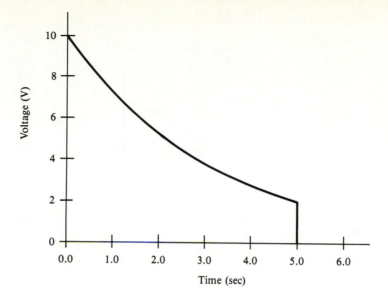

FIGURE 2P–12

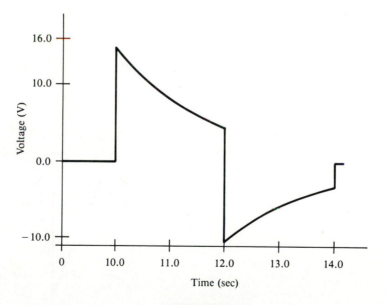

FIGURE 2P–13

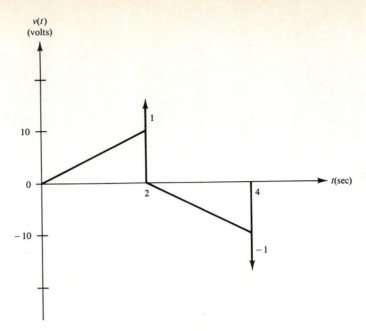

FIGURE 2P–14

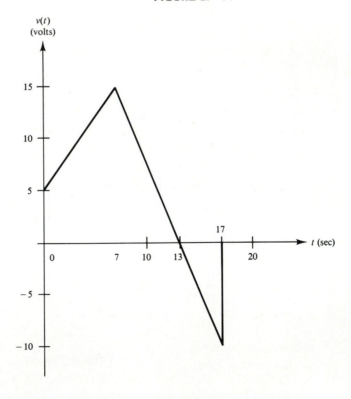

FIGURE 2P–15

CHAPTER 3

Differential Equations

3.1 OBJECTIVES

The major objective of this chapter is to give the student some general concepts that will increase his or her understanding of the nature of differential equations. On completion of this chapter, you should be able to:

- Write differential equations for electrical circuits that can be solved for either voltages or currents.
- Write the characteristic equation for a circuit.
- State the meaning and significance of the natural response and the forced response in a transient solution of a circuit.
- Determine the initial conditions and steady-state response of a circuit.
- Use your knowledge of the initial and final states of a circuit to partially check the results of a transient solution.
- Write differential equations for mechanical systems that can be solved for either velocities or forces.

3.2 INTRODUCTION

In Chapter 1, the relationships between voltage and current in various circuit elements were shown to involve derivatives and integrals. Thus, in solving for transient responses of circuits, the network equations will be integro-differential equations. A number of computer programs are available to solve for the network responses, so they are the preferred method of solution, particularly since several of these can be used on personal computers. However, a knowledge of the general characteristics of the differential equations of networks and their solutions gives insight into the fundamentals of transient responses in electrical circuits.

3.3 DIFFERENTIAL EQUATIONS OF ELECTRICAL CIRCUITS _____

The relationships between current and voltage for the three basic circuit components that were given in Chapter 1 can be used to develop the differential equations of passive circuits. Because the relationships involve both integrals and derivatives, the initial equations may be integro-differential equations. The integrals can be eliminated easily by differentiating each term of the equation.

Figure 3–1 shows a parallel RC circuit driven by an ideal current source. If we want to find the voltage across the parallel elements, we can first apply Kirchhoff's current law to get

$$i_R + i_C = i_S \tag{3-1}$$

We can then write expressions for the currents through the resistor and capacitor using Equations (1–30) and (1–38). Substituting these into Equation (3–1), we obtain

$$\frac{v}{R} + C\frac{dv}{dt} = i_S \tag{3-2}$$

The highest order derivative is of the first order, so this is a first-order differential equation. The dependent variable v and its derivative dv/dt are of the first power and there are no products of different order derivatives. Therefore, the equation is a linear differential equation. The coefficients of v and dv/dt are constants, so we classify the equation as a *first-order linear differential equation with constant coefficients*.

If the term on the right side of such a differential equation is zero, we call the equation a *homogeneous* differential equation. Equation (3–2) is not homogeneous due to the presence of the source current i_S on the right side. When the equation is derived from an electric circuit, we often call the term on the right the *forcing function* or *excitation*. A homogeneous equation describes a circuit without excitation of any sort.

Figure 3–2 shows a series RC circuit driven by an ideal voltage source. We can apply Kirchhoff's voltage law to get

$$v_R + v_C = e_S \tag{3-3}$$

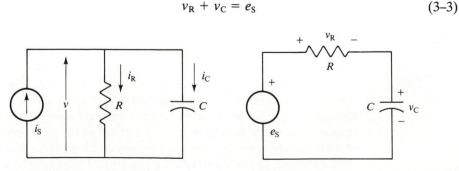

FIGURE 3–1 Parallel RC circuit   **FIGURE 3–2** Series RC circuit

Equations (1–29) and (1–37) give us expressions we can substitute for the voltage drops in Equation (3–3). Doing so, we obtain

$$Ri + \frac{1}{C}\int i\,dt = e_S \tag{3–4}$$

This equation is called an *integral equation.* If we differentiate all three terms, we get the *differential equation*

$$R\frac{di}{dt} + \frac{i}{C} = \frac{de_S}{dt} \tag{3–5}$$

Again, we have a linear first-order differential equation with constant coefficients. If e_S is a constant, its slope, de_S/dt, is zero, so we have a homogeneous equation.

Figure 3–3 shows a series RLC circuit. When we apply Kirchhoff's voltage law to this circuit, we get the equation

$$L\frac{di}{dt} + Ri + \frac{1}{C}\int i\,dt = e_S \tag{3–6}$$

Equation (3–6) is an *integro-differential* equation in which each term represents a voltage. If we differentiate all the terms, we get the differential equation

$$L\frac{d^2 i}{dt^2} + R\frac{di}{dt} + \frac{1}{C}i = \frac{de_S}{dt} \tag{3–7}$$

Equation (3–7) is a second-order linear differential equation with constant coefficients. Again, if e_S is a constant, the equation is homogeneous and can easily be solved for the unknown current.

It is also possible to write an integro-differential equation for a parallel RLC circuit using Kirchhoff's current law. We can develop terms for the current through each component using Equations (1–30), (1–38), and (1–57). When the three terms

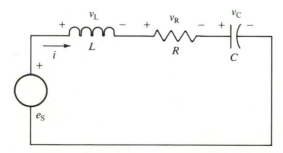

FIGURE 3–3 Series RLC circuit

are added and the equation is differentiated, we obtain a differential equation that can be solved for the voltage across the parallel components:

$$\frac{di_S}{dt} = C\frac{d^2v}{dt^2} + \frac{1}{R}\frac{dv}{dt} + \frac{v}{L} \tag{3-8}$$

In this equation, i_S is the total current from the source.

The order of a differential equation is equal to the number of independent energy storage elements in the circuit. Independent elements are those that cannot be combined in any way to reduce the total number of elements in the circuit. Similar components in series or parallel can easily be combined to give a simpler equivalent circuit.

3.4 FORCED AND NATURAL RESPONSES

The complete solution of a differential equation involving transients includes two parts: the forced response and the natural response. The complete solution is the sum of these two. If the variable being solved for is x, the complete solution is

$$x(t) = x_F(t) + x_N(t) \tag{3-9}$$

where $x_F(t)$ is the forced response and $x_N(t)$ is the natural response.

Methods for finding the forced response with DC and AC excitation are covered in introductory circuit analysis courses. In transient solutions, the forced response is the part of the response whose general character is determined by the excitation applied to the circuit. As a result, it is closely related to the excitation. When AC is applied, the forced response is an AC of the same frequency as, but perhaps a different amplitude and phase shift from, the excitation. With DC excitation, the forced response is a DC of a different amplitude.

When the excitation is continuous AC or DC, the forced response is the response that continues a long period of time after the circuit is first energized, after all transient effects have disappeared. This is sometimes called the *steady-state response*. If the excitation is a transient such as a decaying exponential function, the steady-state response is zero. The forced response will be an exponential function with the same time constant as the excitation.

In most cases, if the circuit includes capacitors and/or inductors, they will not at first hold the energy that they will eventually store as a result of the excitation. The relation between energy and power was given in Equation (1–14) as

$$p = \frac{dw}{dt}$$

Integrating both sides of this equation, we get

$$w = \int p \, dt + C \tag{3-10}$$

Here, the constant C is the energy stored in various circuit components at the beginning of the integration period. As a rule, the stored energy will be different from this initial amount a long time after the start of the transient. The effect of the transient is to cause this change in energy. As indicated by the integral in Equation (3–10), the only way the energy can be changed instantaneously is to provide infinite power. Since infinite power is not available, energy must flow into or out of the components over a period of time to provide any change in energy. The natural response provides the energy required for the circuit to go from its initial state to its forced response. Therefore, the general character of the natural response depends on the circuit component values and configuration.

3.5 THE COMPLETE RESPONSE

If we have a first-order differential equation such as

$$A \frac{dx}{dt} + Bx = C \tag{3–11}$$

where C is the constant forcing function, and the complete response is the sum

$$x(t) = x_F(t) + x_N(t)$$

as given in Equation (3–9), we can rewrite the differential equation as

$$A \frac{d}{dt}(x_F + x_N) + B(x_F + x_N) = C \tag{3–12}$$

With a DC forcing function, the capacitors and inductors will eventually be energized, completing the transient. Thus, the natural response becomes zero after a period of time. At that time, the forced response must satisfy the differential equation; so if t is large enough,

$$A \frac{dx_F}{dt} + Bx_F = C \tag{3–13}$$

Subtracting Equation (3–13) from Equation (3–12), we have the homogeneous differential equation

$$A \frac{dx_N}{dt} + Bx_N = 0 \tag{3–14}$$

Solving for x_N, we get

$$x_N = -\frac{A}{B} \frac{dx_N}{dt} \tag{3–15}$$

One solution of this simple differential equation is

$$x_N = 0 \tag{3-16}$$

but this obviously cannot provide a transition from the initial conditions to the forced response, so it is a trivial solution.

Inspection of a table of derivatives indicates that the derivative of an exponential is an exponential with the same exponent:

$$\frac{d\epsilon^{ax}}{dx} = a\epsilon^{ax} \tag{3-17}$$

If $a = -A/B$, this gives, as a solution of Equation (3–13),

$$x_N = C\epsilon^{-(A/B)t} \tag{3-18}$$

The natural response is thus an exponential, the exponent of which is determined by the circuit components. Its magnitude, since it must cause a transition from the initial state to the forced response, is determined by the magnitudes of the initial state and the forced response.

To simplify the finding of the natural response, the concept of the *characteristic equation* has been developed. To find the characteristic equation from the differential equation, we replace each derivative by s raised to a power equal to the order of the derivative and replace the driving function by zero. The result of applying this procedure to Equation (3–11) is

$$As + B = 0 \tag{3-19}$$

Solving for s, we get

$$s = -\frac{A}{B} \tag{3-20}$$

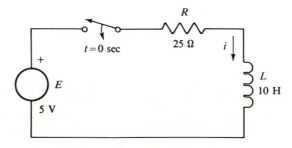

FIGURE 3–4 Circuit for Example 3–1

So the natural response can be written as

$$x_N = C\epsilon^{-A/Bt} \tag{3–21}$$

where the constant C has to be determined. The characteristic equation provides a simple method for finding the general characteristics of the natural response, as we shall see in the following example.

EXAMPLE 3–1   In Figure 3–4, the switch is closed at $t = 0$ seconds. To solve for the current, we can apply Kirchhoff's voltage law and get

$$L\frac{di}{dt} + Ri = E$$

Substituting the specific values for the component sizes and the forcing function, we obtain

$$10\frac{di}{dt} + 25i = 5$$

The characteristic equation is therefore

$$10s + 25 = 0$$

which can be solved for s:

$$s = -2.5$$

Observe that the magnitude of s is R/L, the reciprocal of the time constant. This indicates that the characteristic equation can be solved to find the general character of the natural response. The forcing function in the differential equation was replaced by zero in developing the characteristic equation. The terms that are left are all determined by the values of the components in the circuit. The result is that the general character of the natural response depends on the characteristic equation, and the latter is a function of the circuit components alone.

If we substitute -2.5 for s in Equation (3–21), we get the natural response:

$$i_N = C\epsilon^{-2.5t}$$

The natural response consists of general exponential terms, one for each independent energy storage component in the circuit. When there are two or more independent energy storage components, the exponential responses may include sine waves. This possibility is indicated by Euler's relation between sinusoids and exponentials given in Equation (2–12).

3.6   THE STEADY-STATE DC CIRCUIT

In many transient problems, the circuit is in a steady state before a change occurs in either the excitation or the circuit itself. A long time after the transient starts, the circuit will be in another steady state.

The initial capacitor voltages and inductor currents are called the *initial conditions,* because they determine the energy stored at the start of the transient. The initial conditions must be determined by a steady-state analysis of the circuit just before the change occurs.

With AC excitation, standard circuit analysis techniques can be used. The expressions that give the capacitor voltages and inductor currents just before the transient are written and then evaluated at the time the transient occurs.

Direct current can be defined as a special case of alternating current with zero frequency. The impedances of capacitors and inductors to DC can easily be found using equations developed in Chapter 1.

From Equation (1–43), for capacitors,

$$X_C = \left(\frac{1}{\omega C}\right) \angle -90° \ \Omega$$

If we substitute zero for ω, we get, for the magnitude of the capacitive reactance,

$$|X_C| = \frac{1}{0} = \infty \ \Omega \tag{3–22}$$

Thus, a capacitor has infinite impedance to direct current. Once it is fully charged, no current can flow through it, just as if it had been replaced by an open circuit. Therefore, a fully charged capacitor can be replaced by an open circuit with no change in currents and voltages when the excitation is DC.

The impedance of an inductor can be found from Equation (1–56):

$$X_L = \omega L \angle 90° \ \Omega$$

For DC,

$$|X_L| = 0 \ \Omega \tag{3–23}$$

Thus, under steady-state conditions, just like a short circuit, an inductor provides no opposition to the flow of direct current. It can therefore be replaced by a short circuit without affecting currents and voltages elsewhere in the circuit.

To find the equivalent DC circuit just before the transient starts, we must replace all capacitors by open circuits and all inductors by short circuits. This is illustrated in the following example.

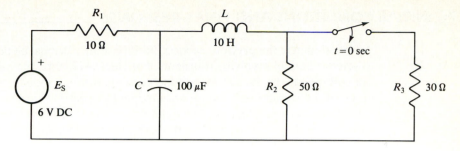

FIGURE 3–5 Circuit for Examples 3–2 and 3–3

EXAMPLE 3–2 Figure 3–5 shows a circuit with DC excitation. At $t = 0$ seconds, the switch is closed, causing a transient response. Find the voltage on the capacitor and the current through the inductor just as the switch is about to close.

We can replace the capacitor by an open circuit and the inductor by a short circuit to get the steady-state circuit shown in Figure 3–6. The switch is not closed until later, so R_3 in Figure 3–5 is not part of the steady-state circuit. In this case, the DC steady-state circuit can be seen to be a simple series circuit.

$V_c(0^+)$, the initial voltage across the capacitor, is found using the voltage divider rule:

$$V_c(0^+) = \frac{R_2}{R_1 + R_2} E_s = \frac{50}{60}(6) = 5 \text{ V}$$

$i_L(0^+)$, the initial current though the inductor, is found using Ohm's law:

$$i_L(0^+) = \frac{E_s}{R_1 + R_2} = \frac{6}{60} = 0.1 \text{ A}$$

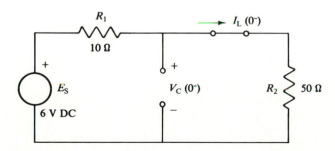

FIGURE 3–6 Initial conditions circuit for Examples 3–2 and 3–3

3.7   INITIAL CONDITIONS AND FINAL RESPONSE

As mentioned in the previous section, initial conditions must be determined in order to solve equations involving transients. For electric circuits, the initial conditions are the voltages across capacitors and the currents through inductors. There will be as many initial conditions as there are independent energy storage elements.

In the circuit illustrated in Figure 3–5, there are two initial conditions, the capacitor voltage and the current through the inductor. As discussed in Section 3.4, we cannot change either the voltage across a capacitor or the current through an inductor instantaneously. An infinitesimal length of time after the switch is closed, the capacitor voltages and inductor currents will be the same as they were just before the switch was closed. These values are called the initial conditions of the circuit. In many cases, some or all initial conditions are zero. The initial conditions are the initial values of the voltages and currents in the circuits that cannot be instantaneously changed.

To deal with initial conditions, we call the voltage just before zero time $v(0^-)$ and the voltage an infinitesimal time later $v(0^+)$. Using this method, we can express the unchanging capacitor voltage at time zero as

$$v_C(0^-) = v_C(0^+) \qquad\qquad (3\text{–}24)$$

Similarly, the inductor voltage is

$$i_L(0^-) = i_L(0^+) \qquad\qquad (3\text{–}25)$$

EXAMPLE 3–3   Find the initial conditions and final response of the circuit shown in Figure 3–5. The initial conditions circuit is shown in Figure 3–6. Here, we have replaced the inductor by a short circuit and the capacitor by an open circuit. The switch will be open, so we may omit both it and R_3 for simplicity. The result is a simple series circuit. The current through the inductor just before closing the switch is

$$i_L(0^-) = \frac{E}{R_1 + R_2} = \frac{6}{60} = 100 \text{ mA}$$

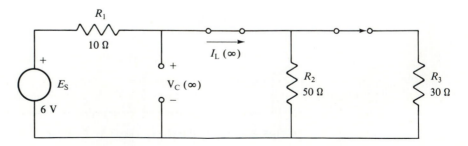

FIGURE 3–7   Final (forced) response circuit for Example 3–3

The current will be the same just after closure, so

$$i_L(0^+) = 100 \text{ mA}$$

The voltage across the capacitor just before closure of the switch will be the drop across R_2, that is,

$$v_C(0^-) = i_L(0^-)R_2 = 0.100(50) = 5 \text{ V}$$

There can be no instantaneous change in capacitor voltage, so

$$v_C(0^+) = 5 \text{ V}$$

If we perform the same series of steps and calculations on the circuit after the switch is thrown, we can find the final response of the circuit. For the circuit shown, closing the switch places R_3 in parallel with R_2. The pair of resistors is then in series with R_1. The equivalent circuit after the transient dies away is shown in Figure 3–7. The resistance of the parallel pair is

$$R_2 \| R_3 = \frac{R_2 R_3}{R_2 + R_3} = \frac{50(30)}{50 + 30}$$

$$= 18.75 \ \Omega$$

The final current through the inductor is

$$i_L(\infty) = \frac{E}{R_1 + R_2 \| R_3} = \frac{6}{10 + 18.75}$$

$$= 208.7 \text{ mA}$$

$R_2 \| R_3$ represents the resistance of R_2 and R_3 in parallel. The forced voltage across the capacitor is the drop across the parallel pair:

$$v_C(\infty) = i_L(\infty)(R_2 \| R_3) = (0.2087)(18.75)$$

$$= 3.913 \text{ V}$$

3.8   SOLUTION OF A FIRST-ORDER DIFFERENTIAL EQUATION _____

The solution of a first-order differential equation is best shown by an example.

EXAMPLE 3–4   The transient response of the circuit shown in Figure 3–8 can be found using a first-order differential equation. We can assume that the initial condition $i_L(0^+)$ is zero. If we need to find the voltage across the parallel components, we can first apply Kirchhoff's current law at the junction between the resistor and the inductor:

$$i_L + i_R = I$$

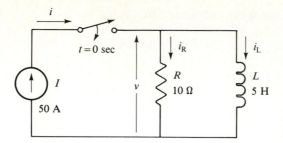

FIGURE 3–8   First-order circuit for Example 3–4

Using equations for the currents given in Chapter 1, we have

$$\frac{1}{L}\int_0^t v\, dt + \frac{v}{R} = I$$

Substituting values from Figure 3–8, we find that

$$\frac{1}{5}\int_0^t v\, dt + \frac{v}{10} = 50$$

Differentiating each term, we obtain

$$\frac{v}{5} + \frac{1}{10}\frac{dv}{dt} = 0$$

This is a homogeneous linear first-order differential equation with constant coefficients that we can solve for the circuit voltages.

The solution of the equation will be of the form

$$v(t) = v_F(t) + v_N(t) \qquad (3\text{--}26)$$

The natural solution can be found first. To do so, we write the characteristic equation:

$$\frac{1}{10}s + \frac{1}{5} = 0$$

Solving for s, we get

$$s = -0.5$$

Substituting for s and x in Equation (3–21), we obtain

$$v_N = C\epsilon^{-0.5t}$$

where C is unknown.

When the excitation is a constant amplitude DC or AC, we can find the forced response by standard circuit analysis techniques. Another method that will work with all types of excitation is the *method of undetermined coefficients*. Since our objective is to provide general knowledge about differential equations rather than to offer complete instructions for older methods of solution, we will use the standard circuit analysis technique.

The source is a DC source, so we can replace the inductor by a short circuit. The switch is closed at $t = 0$ seconds. It can be readily seen that equivalent circuit for the forced response is that shown in Figure 3–9. The resistor is short-circuited by the inductor, so the voltage across the resistor and inductor will be 0 volts. That is,

$$v_F = 0 \text{ V}$$

The complete solution is the sum of the natural and forced solutions:

$$v(t) = v_N + v_F = C\epsilon^{-0.5t} + 0 = C\epsilon^{-0.5t}$$

We now can find the value of the coefficient C. It should satisfy Equation (3–26) at all times. When t is infinite, $v(t)$ is zero, as mentioned in regard to Figure 3–9. This fact does not help us. When $t = 0^-$, the switch is open, so there is no current flowing through the inductor. All of the source current I flows through the resistor at the instant the switch is closed. Since the current through the inductor cannot change instantaneously, the initial condition must be

$$v(0^+) = IR = 50(10) = 500 \text{ V}$$

Evaluating $v(t)$ at $t = 0^+$ seconds, we find that

$$500 = C\epsilon^0$$

so that

$$C = 500$$

The expression for the voltage across the parallel inductor and resistor is, then,

$$v(t) = 500\epsilon^{-0.5t}$$

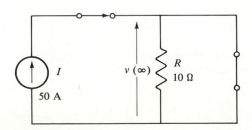

FIGURE 3–9   Forced response circuit for Example 3–4

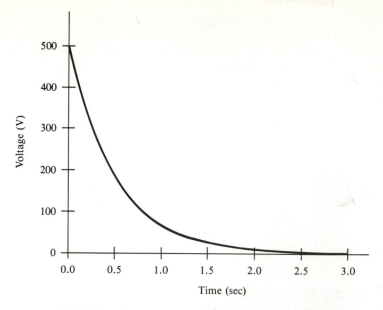

FIGURE 3–10   RL circuit response for Example 3–4

This has a value of 500 volts when the switch is closed, an exponential decay approaching zero an infinite time later, and a time constant of $R/L = 2$. The transient response is illustrated in Figure 3–10.

3.9   SOLUTION OF SECOND-ORDER DIFFERENTIAL EQUATIONS

As with a first-order differential equation, the method of solution of a second-order differential equation is best shown by an example.

EXAMPLE 3–5   Figure 3–11 shows a series RLC circuit energized by a DC voltage source when the switch is closed at $t = 0$ seconds. Find the current in the circuit after the switch is closed.

If we apply Kirchhoff's voltage law, we get

$$L\frac{di}{dt} + Ri + \frac{1}{C}\int i\,dt = E \tag{3–27}$$

Note that we applied Kirchhoff's voltage law to get an equation we could solve for the current. Differentiating each term gives

$$L\frac{d^2 i}{dt^2} + R\frac{di}{dt} + \frac{1}{C}i = 0$$

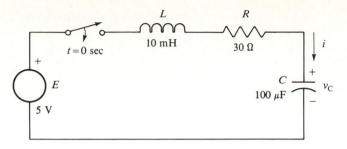

FIGURE 3–11   RLC circuit for Example 3–5

a second-order differential equation that can be solved for the current. To make later calculations simpler, we often divide through by L, to get a coefficient of 1 for the highest order derivative:

$$\frac{d^2 i}{dt^2} + \frac{R}{L}\frac{di}{dt} + \frac{i}{LC} = 0$$

Substituting component values into this equation, we obtain

$$\frac{d^2 i}{dt^2} + (3 \times 10^3)\frac{di}{dt} + (1 \times 10^6)i = 0$$

The circuit has two energy storage elements, the capacitor and the inductor. The initial conditions are the current through the inductor and the voltage across the capacitor at the time the switch is closed. Inspection of Figure 3–11 shows us that the open switch prevents current flow, so

$$i_L(0^-) = i_L(0^+) = 0 \text{ A}$$

There is no way for the capacitor charge to change when the switch is open. Nothing in the circuit diagram can determine the initial capacitor voltage, but let us say that it is given as

$$v_C(0^-) = v_C(0^+) = 0 \text{ V}$$

As with the parallel RL circuit, we can find the natural response from the characteristic equation. This is developed from the differential equation and is

$$s^2 + \left(\frac{R}{L}\right)s + \frac{1}{LC} = 0$$

After entering the values of the components, we have

$$s^2 + 3 \times 10^3 s + 1 \times 10^6 = 0$$

We can find the roots of this quadratic equation using the following form of a familiar formula:

$$s_1, s_2 = -\frac{b}{2} \pm \left[\left(\frac{b}{2}\right)^2 - c\right]^{1/2} \tag{3–28}$$

This formula gives the roots when the s^2 term has a coefficient of unity, b is the coefficient of the second term, and c is the coefficient of the third term. Substituting the values of the coefficients into Equation (3–28), we get

$$s_1, s_2 = -1.5 \times 10^3 \pm [(1.5 \times 10^3)^2 - 1 \times 10^6]^{1/2}$$
$$= -2.618 \times 10^3, -382.0$$

So the characteristic equation can be factored into the product of two terms, and we have

$$(s + 2.618 \times 10^3)(s + 382.0) = 0$$

There will be two exponential terms in the natural solution, one due to each term of the characteristic equation. Again, as in the RL circuit solution, the time constants are determined by the constants in the characteristic equation. We do not yet know the magnitude of either term in the result. The natural solution is

$$i_N(t) = A\epsilon^{-382.0t} + B\epsilon^{-2618t}$$

To find the forced solution, we need to consider the steady-state circuit, shown in Figure 3–12. It can readily be seen that the steady-state current is 0 A, because the capacitor is effectively an open circuit when it is fully charged. So the complete solution is

$$i(t) = i_N(t) + i_F(t) = i_N(t) + 0$$
$$= A\epsilon^{-382.0t} + B\epsilon^{-2618t} \tag{3–29}$$

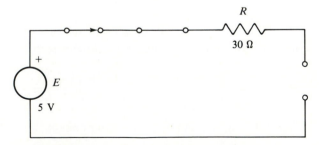

FIGURE 3–12   Steady-state circuit for Example 3–5

We still need to evaluate the coefficients A and B. We can use the two initial conditions, $i_L(0^+)$ and $v_C(0^+)$, to do this. In the series circuit, the current i flows through all elements, so just after the switch is thrown, we have

$$i_L(0^+) = i(0^+) = A\epsilon^0 + B\epsilon^0$$

and this gives us

$$0 = A + B \qquad (3\text{–}30)$$

We can get a second equation involving A and B by differentiating Equation (3–29):

$$\frac{di}{dt} = -382.0A\epsilon^{-382.0t} - 2618B\epsilon^{-2618t} \qquad (3\text{–}31)$$

The second initial condition is the voltage across the capacitor, which is the third term in Equation (3–27). Substituting the initial condition into Equation (3–27) and evaluating it at $t = 0^+$, we get

$$L\frac{di}{dt}\bigg|_{t\,=\,0^+} + Ri(0^+) + v_C(0^+) = E$$

Substituting the given values and initial conditions into this equation and solving for the derivative, we have

$$\frac{di}{dt}\bigg|_{t\,=\,0^+} = \frac{5}{10 \times 10^{-3}} = 500 \text{ A/s}$$

Substituting this into Equation (3–31), we obtain

$$500 = -382.0A - 2618B \qquad (3\text{–}32)$$

Solving simultaneous Equations (3–30) and (3–32), we get

$$A = 0.2236$$

and

$$B = -0.2236$$

Substituting these into Equation (3–27), we find that the complete solution for the current is

$$i(t) = 0.2236\epsilon^{-382.0t} - 0.2236\epsilon^{-2618t}$$

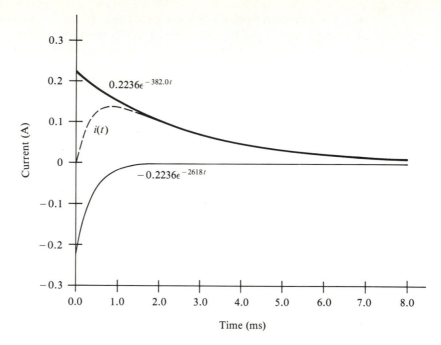

FIGURE 3–13 *Series RLC circuit and response for Example 3–5*

This current is illustrated in Figure 3–13, along with the two exponential components that are added to get the response.

3.10 TYPICAL SOLUTIONS OF DIFFERENTIAL EQUATIONS

If, in Equation (3–28), the quantity $[(b/2)^2 - c]$ under the radical is negative, its square root is imaginary. As a result, the roots of a quadratic characteristic equation may be imaginary or complex. Inspection of Equation (3–28) shows that imaginary and complex roots will occur in complex conjugate pairs, one having a positive imaginary part and the other having a negative imaginary part.

The presence of complex roots means that a circuit's response involves complex exponentials. Euler's equation implies that the natural response for such a network may include sinusoidal components. In fact, if the roots are imaginary, the response includes a sine wave. If the roots are complex, the response includes an exponentially decaying or growing sine wave.

If the quantity under the radical is zero, there are two equal roots. We will see in Chapter 5 that the natural response for such a system is the sum of two terms, an exponential and an exponential multiplied by t.

Much more complicated networks than the two discussed in Sections 3.8 and 3.9 are possible. A general idea of the character of the response in complicated networks can be gained by factoring the characteristic equation. Any factor with a real root represents an exponential term, while a repeated root indicates the sum of an exponential multiplied by t and, perhaps, an exponential.

EXAMPLE 3–6

Consider the third-order differential equation

$$\frac{d^3 i}{dt^3} + 5\frac{d^2 i}{dt^2} + 20\frac{di}{dt} + 16 = 45$$

The characteristic equation is

$$s^3 + 5s^2 + 20s + 16 = 0$$

This can be factored into

$$(s + 1)(s + 2 + j3.464)(s + 2 - j3.464) = 0$$

The three roots include one real root and a complex conjugate pair, so the natural response will include an exponential function and an exponentially decaying sinusoid. Thus, factoring the characteristic equation and inspecting the roots provides information about the general type of natural response for the network. This is discussed in more detail in Section 6.3.5.

As implied earlier in Section 3.3, the order of a differential equation, the degree of its polynomial characteristic equation, and the number of energy storage elements in the circuit or system represented by the differential equation are always equal.

3.11 MECHANICAL DEVICES

Many electrical circuits provide the drive for, or are driven by, mechanical devices. Two simple examples are electric motors and generators. At times, it is possible to ignore the effects of the electrical components on the transient response of such systems. This is so when the response of the electronics is much faster than that of the mechanical devices. It is not always true, however. Generally, as in rotating electrical machinery, the energy is coupled between the mechanical and electrical parts of the system by electromagnetic fields.

In mechanical systems force is required to compress a spring. When the force is removed, the spring will return to its original length. When the spring is compressed, energy is stored in it. Because motion is not involved, we can say that potential energy is stored in the spring. Force is also required to cause a mass to start moving. When the mass is in motion, it tends to keep moving. We say that the moving mass has kinetic energy. In a mechanical system, energy is lost or dissipated by friction. As in electrical circuits, there are two different types of energy storage components and one type of element that dissipates energy.

In a spring, the relation between the velocity of the free end and the resulting force at the fixed end is

$$f = K \int v \, dt \tag{3-33}$$

where f is the force in newtons; v is the relative velocity of the ends of the spring in meters per second or the rate of compression or stretching of the spring; and K is the spring constant, a measure of the strength of the spring. The relationship between the rate of stretching a spring and the force needed to stretch it is shown in Figure 3–14(a). The velocity is that of the right end of the spring. The force, sometimes called the reactive force, is that necessary to keep the left end fixed.

Energy dissipation in mechanical devices may come from sliding friction between a fixed body and a moving part, or it may be provided by a friction device called a *dashpot*, which is illustrated, along with the associated velocity and reactive force vectors, in Figure 3–14(b). Sliding friction is usually nonlinear, so we will not consider it in this book. A dashpot has a piston in a cylinder arranged so that the viscous liquid in the cylinder must pass through a small hole to get from one side of the piston to the other. This provides a reasonable amount of viscous friction. An automotive shock absorber is an example of a dashpot. The force required to move the piston is proportional to the speed of motion; that is,

$$f = Bv = B\frac{dx}{dt} \tag{3-34}$$

where B is the viscous friction coefficient of the device and x is the relative displacement of the end of the dashpot in meters.

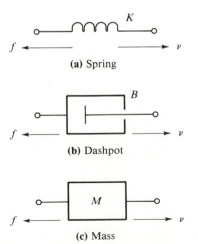

(a) Spring

(b) Dashpot

(c) Mass

FIGURE 3–14 Dynamics of mechanical components

Kinetic energy is stored in moving masses. Figure 3–14(c) shows the relationship between a mass, the reactive force required to move it, and the velocity with which it moves. The basic relation between motion and force is

$$f = Ma = M\frac{dv}{dt} = M\frac{d^2x}{dt^2} \tag{3–35}$$

where M is the mass in kilograms and a is its acceleration in meters per second. Equations (3–33) through (3–35) give the relationship between velocity and force for the three basic dynamic characteristics of mechanical devices.

The transient characteristics of an electric circuit depend on the energy storage and dissipation characteristics of the circuit components. The rate of change of energy, or power, is the product of voltage and current, the two basic variables involved in the differential equations of a circuit. In a mechanical system, the power is the product of velocity and force, so they are analogous to the basic electrical variables. If we have a group of mechanical devices connected together, we can write differential equations that can be solved to find the transient or vibrational responses of the devices considered as a mechanical system. The components may be connected in two ways that resemble parallel and series connections in electric circuits. In one, the total force is the sum of the individual forces, and in the other, the total displacement is the sum of the individual displacements.

There are two electrical analogies to mechanical systems. The *force-current analogy* is the one that should be used if there is magnetic coupling between the electrical and mechanical parts of a system. This is because the current that is the output from an electrical circuit produces the magnetic flux that provides the mechanical force in an electromagnet. This analogy is illustrated in Table 3–1. The other analogy is the *force-voltage analogy*. This should be used if coupling between the electrical and mechanical parts of the system is accomplished by electric fields. This is not used in practical systems, and there is no other reason to use this analogy, so we will not cover it.

TABLE 3–1
Force-Current
Electrical-
Mechanical
Analogy

Electrical	*Mechanical*
Current	Force
$i = C\dfrac{dv}{dt}$	$f = Ma = M\dfrac{dv}{dt} = M\dfrac{d^2x}{dt^2}$
$i = \dfrac{v}{R}$	$f = Bv = B\dfrac{dx}{dt}$
$i = \dfrac{1}{L}\int v\,dt$	$f = K\int v\,dt = Kx$
voltage, v	velocity, v
flux linkage, ϕ	displacement, x
inductance, L	compliance, $1/K$
capacitance, C	mass, M
conductance, $1/R$	friction, B

Figure 3–15 is a schematic drawing of a mass, spring, and dashpot connected so that all have the same motion. In this type of schematic drawing, the crosshatched area is fixed in position, and the common points on the linkages between all the devices all move the same amount. Unfortunately, the diagram is not as easy to interpret as a circuit schematic. The mass moves with the top of both the spring and the dashpot. Since the other sides of the dashpot and the spring are fixed, the displacements of all three elements of the system are the same.

The situation is similar to that found in basic series or parallel electric circuits. In applying Kirchhoff's laws to those circuits, we add the variable that is not the same in all elements. In parallel circuits, the voltages are the same on each element, so we add the currents. In series currents, the voltages may be different, so we add them in setting up the equation of the network.

EXAMPLE
3–7
In Figure 3–15, the displacement for all three elements is the same, so we must sum the forces. The sum is equal to the total force f_T. That is,

$$f_T = \Sigma f_i$$

Entering the specific values of the individual forces required for the motion in each component, we obtain

$$f_T = f_M + f_B + f_K \tag{3–36}$$

This equation resembles one of Kirchhoff's laws. Since we are using the force-current analogy, we can say that it is the analog of Kirchhoff's current law. That means that this system is analogous to a parallel electric circuit.

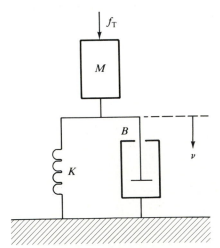

FIGURE 3–15   Mechanical system for Example 3–7

If we substitute into Equation (3–36) the expressions for the forces required to move a mass and the piston of a dashpot and to compress a spring from Equations (3–33) through (3–35), we have

$$f_T = M\frac{dv}{dt} + Pv + K\int vdt$$

Differentiating each term, we get the differential equation of the system:

$$\frac{df_T}{dt} = M\frac{d^2v}{dt^2} + B\frac{dv}{dt} + Kv$$

Figure 3–16(a) shows a system in which the velocities of the components are different. There are only two velocities identified in the figure. The significant velocities for the spring and the dashpot are the relative velocities of the ends of the two elements. The velocity for the mass is that of its center of gravity.

EXAMPLE 3–8 Write the differential equation for the system in Figure 3–16(a).

In this system, there are two velocities: v_2, the velocity of the mass, and v_1, the velocity of the junction between the spring and the dashpot. The velocity causing

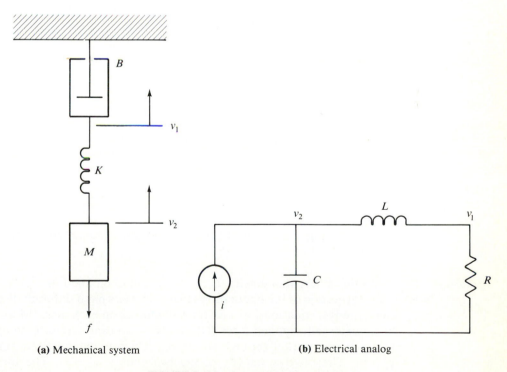

(a) Mechanical system   (b) Electrical analog

FIGURE 3–16 Systems for Example 3–8

the spring to be stretched is $v_1 - v_2$. The external force f applied to the mass must equal the force due to its acceleration plus the component due to the stretching of the spring.

From Equations (3–35) and (3–33), an equation for the forces acting on the mass is

$$f = M\frac{dv_2}{dt} + K \int (v_2 - v_1)dt \qquad (3\text{–}37)$$

Inspection of Figure 3–16(a) shows that if v_2 is larger than v_1, the force on the spring should act in the same direction as the force producing acceleration in the mass.

From Equations (3–34) and (3–33), another equation for the forces at the junction between the spring and the dashpot is

$$K \int (v_1 - v_2)dt - Bv_1 = 0 \qquad (3\text{–}38)$$

If v_1 is larger than v_2, the spring will be stretched as the dashpot is compressed, so the signs of the two terms will be different.

We now have two differential equations in two unknowns, v_1 and v_2. These can be solved readily by the use of Laplace transforms. Using Table 3–1, we can develop an electric circuit that is analogous to the mechanical system the differential equations describe. The analog of force is current, the analog of mass is capacitance, and the analog of velocity is voltage. For the spring constant K, we have the reciprocal of inductance. Substituting these into Equation (3–37), we have

$$i = C\frac{dv_2}{dt} + \frac{1}{L} \int (v_2 - v_1)dt \qquad (3\text{–}39)$$

Using the analogs for voltage and inductance and the fact that the analog of the coefficient of friction is the reciprocal of resistance, we find that

$$\frac{1}{L} \int (v_2 - v_1)dt - \frac{v_1}{R} = 0 \qquad (3\text{–}40)$$

Equations (3–39) and (3–40) are the nodal equations for the parallel RLC circuit shown in Figure 3–16(b).

EXAMPLE 3–9   Write the differential equations of the system diagramed in Figure 3–17.

Inspection of the figure shows that each mass has a different displacement. We can write two equations, one each for the forces on each mass. Mass M_1 is acted on by an externally applied force $f(t)$ and two opposing forces. One of the opposing forces is transmitted through spring K_2, which is extended an amount equal to $x_2 - x_1$. The relative velocity of the two ends is $v_2 - v_1$. The force transmitted through the spring is

$$f_K = K_2 \int (v_2 - v_1)dt$$

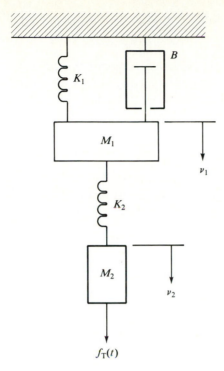

FIGURE 3–17   Complex mechanical system for Example 3–9

The other opposing force is caused by the inertia of the mass and is equal to $M_2\,dv_2/dt$. The applied force must equal the sum of these two forces:

$$f_\mathrm{T} = M_2\frac{dv_2}{dt} + K_2 \int (v_2 - v_1)dt$$

The forces on mass M_1 are those provided by its inertia, by spring K_1, and by the dashpot, as well as an opposing force transmitted through spring K_2. Summing these up, we have

$$0 = K_2 \int (v_1 - v_2)dt + K_1 \int v_1\,dt + Bv_1 + M_1\frac{dv_1}{dt}$$

We can now differentiate each integro-differential equation to get

$$\frac{df_\mathrm{T}}{dt} = M_2\frac{d^2v_2}{dt^2} + K_2(v_2 - v_1)$$

and

$$0 = K_2(v_1 - v_2) + K_1 v_1 + B\frac{dv_1}{dt} + M_1\frac{d^2v_1}{dt^2}$$

We thus have two simultaneous differential equations that can be solved for the two unknowns.

Like electrical components, mechanical components can store energy. Kinetic energy, the energy of motion, is stored in moving masses. Because the velocity $v(0^-)$ of a moving mass cannot be changed instantaneously without infinite power, we have, for any mass,

$$v(0^-) = v(0^+) \tag{3-41}$$

Potential energy, the energy of position, is stored in a compressed or stretched spring. Because $x(0^-)$, the amount the spring is stretched, cannot be changed instantaneously without infinite power, we have, for any spring,

$$x(0^-) = x(0^+) \tag{3-42}$$

Equations (3–41) and (3–42) illustrate the initial conditions for mechanical devices.

We see, then, that mechanical systems have energy storage characteristics. One parameter, mass, stores kinetic energy; another parameter, the spring constants, reflects stored potential energy. The friction present in the system dissipates energy. Mechanical systems are therefore similar to electrical circuits, so we can expect the same types of transient responses.

PROBLEMS

3–1. For the circuit shown in Figure 3P–1, write a differential equation that can be solved for the current after the switch is closed at 0 seconds.

3–2. Write equations that can be used to find the voltage drops across the resistor and inductor after the current is found in Problem 3–1. Any of these equations may be, but are not necessarily, differential equations.

3–3. Write a differential equation that can be solved for the voltage drop across the source in Figure 3P–2 for the period after the switch is closed at $t = 0$ seconds.

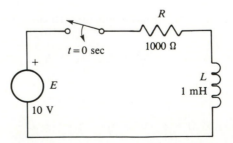

FIGURE 3P–1

3–4. Using the equation found for the voltage drop in Problem 3–3, write equations that can be solved for the currents through the resistor and the inductor.

3–5. Find the differential equation for the voltage drop across the parallel combination of components for the period after the switch is closed at $t = 0$ seconds in Figure 3P–3.

3–6. What are the equations for i_L, i_R, and i_C in Figure 3P–3?

3–7. The switch in the circuit shown in Figure 3P–4 is closed at $t = 0$ seconds. What is the differential equation for the current after switch closure?

3–8. Using the result from Problem 3–7, what are the equations for v_C, v_L, and v_R in the circuit of Figure 3P–4?

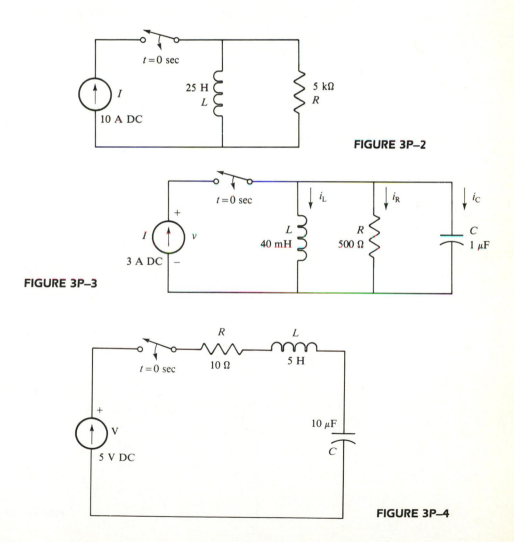

FIGURE 3P–2

FIGURE 3P–3

FIGURE 3P–4

3–9. Write the characteristic equation for the differential equation found in Problem 3–1.

3–10. What is the characteristic equation for the differential equation found in Problem 3–3?

3–11. Give the characteristic equation for the differential equation found in Problem 3–5.

3–12. What is the characteristic equation for the differential equation found in Problem 3–7.

3–13. Find the forced, natural, and complete responses for the current in the circuit shown in Figure 3P–5 after the switch is thrown at $t = 0$ seconds. E is 5 volts DC, and the capacitor is uncharged before the switch is thrown. What are the complete responses for the voltage drops across the capacitor and resistor?

3–14. In the circuit shown in Figure 3P–6, E is 10 volts DC and the switch is closed at $t = 0$ seconds.

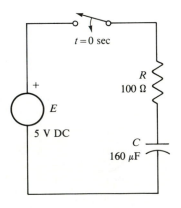

FIGURE 3P–5

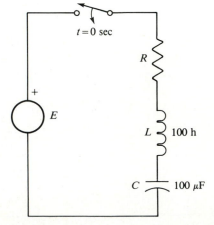

FIGURE 3P–6

   a. If $R = 5\ \Omega$, what is the characteristic equation? What general type of response is found?

   b. What type of response results if $R = 0.1\ \Omega$?

3–15. Using the force-current equivalents in Table 3–1, draw the electric circuit that is analogous to the mechanical system shown in Figure 3P–7. Write the differential equations for both. What type of transient response can be expected?

3–16. Using Table 3–1, draw the electric circuit equivalent for the mechanical system shown in Figure 3P–8. What is the differential equation for the mechanical system?

3–17. Draw the equivalent electric circuit for the system shown in Figure 3P–9. Write the differential equations for the mechanical system.

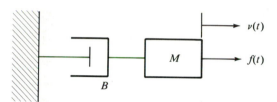

FIGURE 3P–7

FIGURE 3P–8

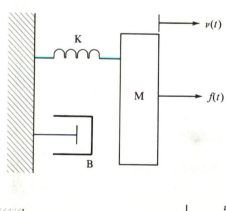

FIGURE 3P–9

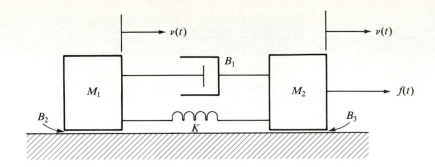

FIGURE 3P–10

3–18. There are two unknown displacements in the system shown in Figure 3P–10. Write the two differential equations that are necessary to analyze the mechanical system. Find an equivalent electric circuit. Note that there is friction between the masses and the surface they are resting on.

CHAPTER 4

Laplace Transforms

4.1 OBJECTIVES

On completion of this chapter, you should be able to solve differential equations through the use of Laplace transforms. In so doing, you should be able to:

- Use a table of transform pairs and a table of operation pairs to find the Laplace transforms of common time domain functions and to perform the inverse operation for common functions in the Laplace transform domain.

- Use the definition of the Laplace transform to find the transforms of less common functions if they cannot be found in more complete tables of transform pairs.

- Use algebraic manipulation to simplify the process of inverting the transforms of the solutions of differential equations.

4.2 INTRODUCTION

Differential equations can be solved by the classical method described in Chapter 3. The method is useful with comparatively simple problems, such as those given in the examples in that chapter. In many cases, the solution of the simpler problems can even be memorized.

For more complex problems, Laplace transforms provide a method of solution that is easier than the classical method. A brief examination of the classical method, however, gives insight into the general characteristics of solutions and into typical solutions of the differential equations describing electric circuits.

A method for solving practical electric circuit transient problems was developed late in the 19th century by the electrical engineer Oliver Heaviside. An operational calculus, this method drew considerable opposition from mathematicians and more conservative engineers because it did not have a proven mathematical foundation. The opposition was justified, because if a method has not been proven mathematically, there is always the chance that it may not work in some application which has not yet been tested.

About a century before Heaviside, the French mathematician Simon Laplace had developed and proved mathematically an operational method that was eventually found to be useful in transient solutions. It is easier to handle initial conditions with the Laplace method than with Heaviside's method. As a result, Laplace transforms have replaced Heaviside's operational calculus in the solution of differential equations of electric circuits. Note, however, that no significant errors have yet been found in Heaviside's solutions since he worked them out.

Numerous transforms have been developed in mathematics in order to simplify calculation. A simple example is logarithms. If we want to multiply two numbers, we can do so by finding their logarithms and adding them. The sum is the logarithm of the product. To get the product itself, we must perform the inverse of the original transformation—that is, find the antilogarithm of the product. If we call the logarithm a transform, the antilogarithm will be the inverse transform. Mathematically,

$$\log xy = \log x + \log y \tag{4–1}$$

This transformation may not seem to save much time or effort, but before the days of hand calculators it was the generally accepted method for performing multiplication simply and accurately. The logarithm thus allows the use of simple mathematics to solve moderately complex mathematical problems.

The Laplace transform converts integro-differential equations into equations that may be solved algebraically.

Other types of transform have been developed. One, the Fourier transform, is used to find the frequency content or spectrum of a wave when the amplitude-versus-time function is known. Because it is not possible to find Fourier transforms for several functions that occur in transient problems, we do not use Fourier transforms in these problems. Laplace and Fourier transform methods are most useful for analog circuits and waveforms; if, by contrast, the forcing functions or waveforms are sampled, as in digital systems, the z-transform is used. The z-transform was developed from the Laplace transform. It can be used to transform the difference equations that describe the operation of sampled data systems into algebraic equations. Sampled data systems are active systems and thus are not within the intended scope of this text. More information may be found in some, but not all, books on control system theory.

4.3   THE LAPLACE TRANSFORM

After finding the Laplace transform of a function of time, we have a function of a new variable s. This variable is called an *operator*. We can express the transformation as

$$\mathcal{L}[f(t)] = F(s) \tag{4–2}$$

In words, the Laplace transform of a function of time $f(t)$ is $F(s)$.

The time function corresponding to an expression in s is called the *inverse transform* and is given by

$$\mathcal{L}^{-1}[F(s)] = f(t) \tag{4–3}$$

Note that we do not say that $f(t)$ equals $F(s)$; it does not. The two expressions are different ways of describing the same thing, just as are the time function for a voltage waveform and the Fourier series describing the frequencies present in the wave.

The definition of the Laplace transform $F(s)$ of a function $f(t)$ is

$$F(s) = \mathcal{L}[f(t)] = \int_0^\infty f(t)\epsilon^{-st}\, dt \tag{4–4}$$

Exponents must be dimensionless. So since t has the dimension of time, the dimension of s must be the reciprocal of time, $1/t$, or frequency. We will see later that s can be complex; that is,

$$s = \sigma + j\omega \ \text{sec}^{-1} \tag{4–5}$$

Here, ω is real frequency; so we say that the variable s is a complex frequency variable.

Note that the integral in Equation (4–4) is a definite integral evaluated over a period of time. The variable t will not appear in the transform $F(s)$. The transform will be a function of s. Later we will see that s is a frequency function. $F(s)$ and $f(t)$ are called a *transform pair*.

4.4 LAPLACE TRANSFORM PAIRS

Table 4–1 lists a number of common functions and their Laplace transforms. Table 4–2 lists several time domain operations or properties and their corresponding Laplace transforms. These short tables, along with algebraic manipulation, can be used to solve the overwhelming majority of transient problems. More complete tables are also available.

The Fourier transform can be considered a special case of the Laplace transform in which the real part of s, σ, is equal to zero. As a result, the Fourier transform is a function of $j\omega$. If we have a table of Fourier transforms, we can use it as a table of Laplace transform pairs by substituting s for $j\omega$ in the Fourier transforms. However, it should be noted that the Fourier transform does not exist for a number of functions that have Laplace transforms.

The function transforms listed in Table 4–1, along with the transforms of operations in Table 4–2, should allow any common functions to be transformed. If any other functions must be transformed, Equation (4–4) can be used to find the transform. The transforms of several possible excitation waveforms will be developed as examples. Other transforms in the table are developed in the appendix.

TABLE 4–1
Laplace
Transform Pairs

Pair	$f(t)$	$F(s)$
1	1	$\dfrac{1}{s}$
2	ϵ^{-at}	$\dfrac{1}{s+a}$
3	$\cos \omega t$	$\dfrac{s}{s^2+\omega^2}$
4	$\sin \omega t$	$\dfrac{\omega}{s^2+\omega^2}$
5	t	$\dfrac{1}{s^2}$
6	$t\epsilon^{-at}$	$\dfrac{1}{(s+a)^2}$
7	$\epsilon^{-at}\cos \omega t$	$\dfrac{s+a}{(s+a)^2+\omega^2}$
8	$\epsilon^{-at}\sin \omega t$	$\dfrac{\omega}{(s+a)^2+\omega^2}$
9	$\delta(t)$	1

Note: All time functions are zero for $t < 0$ seconds. This means that, for pairs 1 through 8, the time domain expression should actually be multiplied by $U(t)$.

TABLE 4–2
Transform
Properties

Pair	Time Domain	Frequency Domain (s-plane)
A	$Af(t)$	$AF(s)$
B	$f_1(t) + f_2(t)$	$F_1(s) + F_2(s)$
C	$d[f(t)]$	$sF(s) - f(0^+)$
D	$\dfrac{d^2}{dt^2}[f(t)]$	$s^2F(s) - sf(0^+) - f'(0^+)$
E	$\displaystyle\int_0^t f(t)\,dt$	$\dfrac{F(s)}{s}$
F	$f(t-a)U(t-a)$	$\epsilon^{-as}F(s)$
G	$\epsilon^{-at}f(t)$	$F(s+a)$

All the time functions in Table 4–1 are defined such that their value is zero for $t < 0$ seconds. A more complete expression for the general excitation function $f(t)$ is $U(t)f(t)$ for pairs 1 through 8 in the table. In most problems, the transient begins at zero seconds. If not, time may usually be redefined to be zero at the start of the transient. Then the transform pairs given in the table may be used.

4.4.1   The Unit Step Function $U(t)$

The unit step function $U(t)$ is commonly used to define the sudden change in voltage or current that often starts a transient response. The function is illustrated in Figure 4–1.

EXAMPLE 4–1

Find the Laplace transform of the unit step function.

Substituting the unit step function into Equation (4–4), we obtain

$$\mathcal{L}[U(t)] = \int_0^\infty U(t)\epsilon^{-st}\,dt \qquad (4\text{–}6)$$

Since the range of integration of the definite integral starts where the step function becomes 1, we may replace the step function by the constant 1 in the integral. We then have

$$\mathcal{L}[U(t)] = \int_0^\infty \epsilon^{-st}\,dt = \frac{\epsilon^{-st}}{-s}\bigg|_0^\infty = \frac{1}{s}, \qquad \text{for } s > 0 \qquad (4\text{–}7)$$

This is transform pair 1 in Table 4–1.

4.4.2   The Unit Impulse Function $\delta(t)$

The unit impulse function is another important special function used in Laplace transform transient analysis. It is sometimes called the *Dirac delta function*. The function is illustrated in Figure 4–2.

EXAMPLE 4–2

Find the Laplace transform of the unit impulse function.

Substituting the impulse function into the defining integral, we get

$$\mathcal{L}[\delta(t)] = \int_0^\infty \delta(t)\epsilon^{-st}\,dt \qquad (4\text{–}8)$$

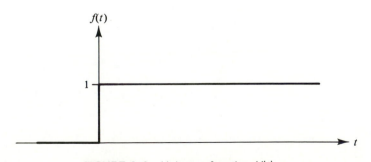

FIGURE 4–1   Unit step function $U(t)$

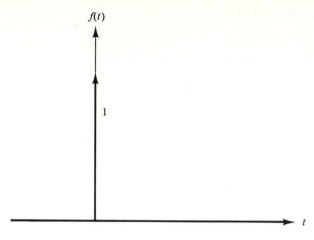

FIGURE 4–2 Unit impulse function $\delta(t)$

The impulse function is zero for $t \neq 0$ seconds, but has a value at $t = 0$ seconds. At that point, the exponential ϵ^{-st} is equal to 1. So the product $\delta(t)\epsilon^{-st}$ will be zero for $t \neq 0$ and equal to $\delta(t)$ at $t = 0$ seconds. Thus, we can simplify the integral to

$$\mathcal{L}[\delta(t)] = \int_0^\infty \delta(t)\, dt \qquad (4\text{–}9)$$

By definition, the integral of the unit impulse function is equal to 1, so

$$\mathcal{L}[\delta(t)] = 1 \qquad (4\text{–}10)$$

This is transform pair 9 in Table 4–1.

4.4.3 The Exponential Function ϵ^{-at}

The decaying exponential function is a common factor in the transient response of simple circuits. The function is shown in Figure 4–3. In analyzing transients, we usually express the exponential function as ϵ^{-at}, but, as mentioned earlier, it is zero for $t < 0$ seconds, so it is really $U(t)\epsilon^{-at}$.

EXAMPLE 4–3 Find the Laplace transform of the exponential function ϵ^{-at}.

Substituting the exponential function into Equation (4–4), we obtain

$$\mathcal{L}[\epsilon^{-at}] = \int_0^\infty \epsilon^{-at}\epsilon^{-st}\, dt = \int_0^\infty \epsilon^{-(s + a)t}\, dt$$

$$= \frac{\epsilon^{-s + a}}{-(s + a)}\Big|_0^\infty = \frac{1}{s + a} \qquad (4\text{–}11)$$

This is transform pair 2 in Table 4–1.

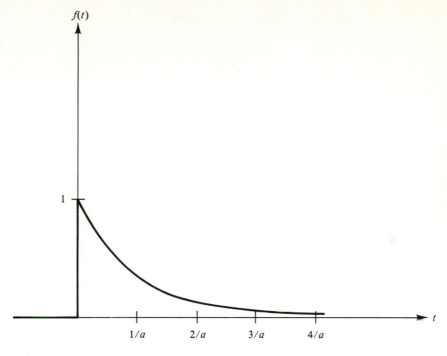

FIGURE 4–3   Switched decaying exponential function ϵ^{-at}

EXAMPLE 4–4   Find the Laplace transform of $f(t) = \epsilon^{-4t}$.

Using transform pair 2 in Table 4–1, we see that $a = 4$; so

$$F[s] = \frac{1}{s + a} = \frac{1}{s + 4}$$

4.4.4   The Cosine Wave cos ω*t*

The cosine function commonly occurs as an excitation function or in a transient response. It is shown in Figure 4–4.

EXAMPLE 4–5   Find the Laplace transform of $\cos \omega t$.

Substituting the cosine wave function into the defining integral, we get

$$\mathcal{L}[\cos \omega t] = \int_0^\infty (\cos \omega t)\epsilon^{-st}\, dt \qquad (4\text{–}12)$$

Using Equation (2–15), we obtain

$$\mathscr{L}[\cos \omega t] = \int_0^\infty \frac{[\epsilon^{-st}(\epsilon^{j\omega t} + \epsilon^{-j\omega t})]}{2} dt$$

$$= \frac{1}{2}\int_0^\infty \epsilon^{j\omega t}\epsilon^{-st} dt + \frac{1}{2}\int_0^\infty \epsilon^{-j\omega t}\epsilon^{-st} dt$$

$$= \frac{1}{2}\mathscr{L}[\epsilon^{j\omega t}] + \frac{1}{2}\mathscr{L}[\epsilon^{-j\omega t}]$$

which, using transform pair 2, yields

$$\mathscr{L}[\cos \omega t] = \frac{1}{2}\left[\frac{1}{s + j\omega} + \frac{1}{s - j\omega}\right] = \frac{s}{s^2 + \omega^2} \qquad (4\text{–}13)$$

This is transform pair 3.

EXAMPLE 4–6   Find the Laplace transform of $f(t) = \cos 10\pi t$.
Using transform pair 3, we have $\omega = 10\pi$; so

$$F[s] = \frac{s}{s^2 + \omega^2} = \frac{s}{s^2 + 100\pi^2}$$

4.4.5   Transforms of Other Common Functions

Transform pairs 4 and 5 are developed in the appendix. Pairs 6 through 8 will be developed in Section 4.5.8. Here, we illustrate their use by several examples.

EXAMPLE 4–7   Find the Laplace transform of $f(t) = \sin 120\pi t$.
Using transform pair 4, we have $\omega = 120\pi$ and

$$F[s] = \frac{\omega}{s^2 + \omega^2} = \frac{120\pi}{s^2 + 14400\pi^2}$$

EXAMPLE 4–8   Find the Laplace transform of $f(t) = t$.
Using transform pair 5, we obtain

$$F[s] = \frac{1}{s^2}$$

EXAMPLE 4–9   Find the Laplace transform of $f(t) = t\epsilon^{-16t}$.
Using transform pair 6 we get

$$F[s] = \frac{1}{(s + 16)^2}$$

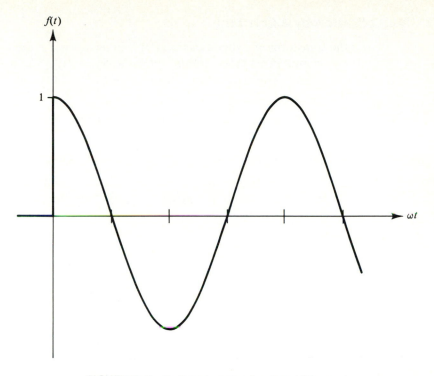

FIGURE 4–4 Switched cosine function $U(t) \cos \omega t$

EXAMPLE 4–10 Find the Laplace transform of the exponentially decaying cosine wave $f(t) = \epsilon^{-2t} \cos 4t$.

Here, we need to use transform pair 7. We obtain

$$F[s] = \frac{s + 2}{(s + 2)^2 + 16}$$

EXAMPLE 4–11 Find the Laplace transform of the function $f(t) = \epsilon^{-3t} \sin 2\pi t$.

Here, transform pair 8 is used to get

$$F[s] = \frac{2\pi}{(s + 3)^2 + 4\pi^2}$$

4.5 OPERATIONAL RELATIONSHIPS

The operational rules listed in Table 4–2 are useful in determining the s-plane equivalent of some time domain functions. We derive these rules in this section.

4.5.1 Multiplication by a Constant

If the transform of a time function $f(t)$ is known, what is the transform of that time function multiplied by a constant [symbolically, $Af(t)$]? The transform can be found by using the definition of the Laplace transform, Equation (4–4). We have:

$$\mathcal{L}[Af(t)] = \int_0^\infty Af(t)\epsilon^{-st}\, dt \qquad (4\text{–}14)$$

Taking the constant outside of the integral sign, we get

$$\mathcal{L}[Af(t)] = A\int_0^\infty f(t)\epsilon^{-st}\, dt \qquad (4\text{–}15)$$

So

$$\mathcal{L}[Af(t)] = A\mathcal{L}[f(t)] = AF(s) \qquad (4\text{–}16)$$

That is, the Laplace transform of $Af(t)$ is A times the Laplace transform of $f(t)$. This relation is expressed as operation pair A in Table 4–2.

4.5.2 The Sum of Two Functions

If we have an expression that is the sum of two functions, i.e.,

$$f(t) = g(t) + h(t) \qquad (4\text{–}17)$$

the associated integral may be separated into two integrals. Substituting into the defining equation for the transform, we have

$$\mathcal{L}[g(t) + h(t)] = \int_0^\infty [g(t) + h(t)]\epsilon^{-st}\, dt$$

$$= \int_0^\infty g(t)\epsilon^{-st}\, dt + \int_0^\infty h(t)\epsilon^{-st}\, dt$$

$$= F_1(s) + F_2(s) \qquad (4\text{–}18)$$

This relation is expressed as pair B in Table 4–2.

4.5.3 The Product of Two Functions

The transform of a product of two functions, say, $g(t)h(t)$, is not in general the product of the transforms of the individual functions. That is, in general,

$$\mathcal{L}[g(t)h(t)] \neq G(s)H(s) \qquad (4\text{–}19)$$

A proof of this statement for electrical variables can be given based on the superposition theorem. [A proof of Equations (4–16) and (4–18) can also be given based on the superposition theorem.]

EXAMPLE 4–12 Show that the Laplace transform of the product $\epsilon^{-at} \sin \omega t$ is not the product of the transform of ϵ^{-at} and the transform of $\sin \omega t$.

The transforms of the individual factors are found using transform pairs 2 and 4 of Table 4–1. Multiplying, we then obtain

$$\mathscr{L}[\epsilon^{-at}] \cdot \mathscr{L}[\sin \omega t] = \left[\frac{1}{s + a}\right] \cdot \left[\frac{\omega}{s^2 + \omega^2}\right] = \frac{\omega}{(s + a)(s^2 + \omega^2)}$$

The transform of the product is given by pair 8 in Table 4–1:

$$\mathscr{L}[\epsilon^{-at} \sin \omega t] = \frac{\omega}{(s + a)^2 + \omega^2} \neq \mathscr{L}[\epsilon^{-at}]\mathscr{L}[\sin \omega t]$$

4.5.4 The Derivative

The first derivative of the unknown variable is usually present in the integro-differential equations of electrical networks. The Laplace transform of the derivative is found using the basic definition of the transform:

$$\mathscr{L}\left[\frac{df(t)}{dt}\right] = \int_0^\infty \left[\frac{df(t)}{dt}\right]\epsilon^{-st}\, dt$$

The integral can be evaluated by parts. Let $u = \epsilon^{-st}$ and $dv = df(t)/dt$. Then $du = -s\epsilon^{-st}\, dt$ and $v = f(t)$. So

$$\mathscr{L}\left[\frac{df(t)}{dt}\right] = (uv)\Big|_0^\infty - \int_0^\infty v\, du$$

$$= \epsilon^{-st} f(t)\Big|_0^\infty - \int_0^\infty f(t)(-s\epsilon^{-st})\, dt \qquad (4\text{–}20)$$

$$= 0 - f(0^+) + s\int_0^\infty f(t)\epsilon^{-st}\, dt$$

Hence,

$$\mathscr{L}\left[\frac{df(t)}{dt}\right] = sF(s) - f(0^+) \qquad (4\text{–}21)$$

This equation is expressed as pair C in Table 4–2. A time derivative can thus be seen to be the equivalent of multiplication by s, an algebraic operation, in the complex frequency plane. The constant $f(0^+)$ is an initial condition. In electric circuits, this could be the initial voltage across a capacitor or the initial current through an inductor.

4.5.5 The Second Derivative

We can use operation pair C of Table 4–2 to find the Laplace transform of higher order derivatives. Substituting the derivative $f'(t)$ for $f(t)$, we have

$$\mathcal{L}\left[\frac{d\{f'(t)\}}{dt}\right] = s\mathcal{L}[f'(t)] - f'(0^+)$$

Replacing $\mathcal{L}[f'(t)]$ by $sF(s) - f(0^+)$, we obtain

$$\mathcal{L}\left[\frac{d^2}{dt^2}\{f(t)\}\right] = s^2 F(s) - sf(0^+) - f'(0^+) \qquad (4\text{--}22)$$

This equation is expressed as operation pair D in Table 4–2. The second derivative in the time domain is represented in the s-plane by the product of s^2 and the transform of the original function. The other two terms can be seen to depend on the initial value of the variable and its rate of change. Consequently, they are initial conditions. The procedure just used can be extended to find the Laplace transforms of higher order derivatives.

4.5.6 The Integral

Often, the first equation we set up for finding transient responses will include an integral. We can use the definition of the Laplace transform to find the transform of the integral:

$$\mathcal{L}\left[\int_0^t f(t)\,dt\right] = \int_0^\infty \left[\int_0^t f(t)\,dt\right]\epsilon^{-st}\,dt$$

We can solve this equation by integrating by parts. If we let

$$u = \int_0^t f(t)\,dt$$

and

$$dv = \epsilon^{-st}\,dt$$

then

$$du = f(t)\,dt$$

and

$$v = -\frac{\epsilon^{-st}}{s}$$

Using the formula for integration by parts, we obtain

$$\int_0^t f(t)\, dt = uv \Big|_0^\infty - \int_0^\infty v\, du$$

$$= -\frac{\epsilon^{-st}}{s} \int_0^t f(t)\, dt \Big|_0^\infty + \int_0^\infty \frac{\epsilon^{-st}}{s} f(t)\, dt$$

The first term on the right side is zero at the upper limit, due to the presence of the exponential term. At its lower limit, the integral of $f(t)$ is zero. So the first term is zero. The second term is the Laplace transform of $f(t)$, $[F(s)]$, divided by s. So

$$\mathcal{L}\left[\int_0^t f(t)\, dt \right] = \frac{F(s)}{s} \qquad (4\text{--}23)$$

This equation is represented by operation pair E of Table 4–2. Thus, in words, the Laplace transform of the definite integral of $f(t)$ is the Laplace transform of $f(t)$, i.e., $F(s)$, divided by s.

4.5.7 The Time Shift

On occasion, it is necessary for some function to start at some time other than zero seconds. If the function is to start, say, at a seconds, we must replace t in its equation by $t - a$. If we are shifting a function that originally was switched on at zero seconds, we must include the unit step function in our equation. For instance, the exponential ϵ^{-bt} can be rewritten $U(t)\epsilon^{-bt}$. If we wish to shift this function so that it starts at $t = a$ seconds, we must shift both the exponential and the step function. That is,

$$f(t - a) = U(t - a)\epsilon^{-b(t - a)} \qquad (4\text{--}24)$$

A time shift or translation has a specific effect on the Laplace transform of a function. We can find out what this effect is by using the defining integral of the Laplace transform:

$$\mathcal{L}[f(t - a)U(t - a)] = \int_0^\infty U(t - a)f(t - a)\epsilon^{-st}\, dt$$

$$= \int_a^\infty f(t - a)\epsilon^{-st}\, dt \qquad (4\text{--}25)$$

The right side can be integrated by a change in the variable of integration. That is, let $x = t - a$. Then if $t = a$, $x = 0$, and if $t = \infty$, $x = \infty$. We then have

$$\mathcal{L}[f(t - a)U(t - a)] = \int_0^\infty f(x)\epsilon^{-s(x + a)}\, dx$$

$$= \epsilon^{-as} \int_0^\infty f(x)\epsilon^{-sx}\, dx \qquad (4\text{--}26)$$

$$= \epsilon^{-as} F(s)$$

This equation is represented by operation pair F in Table 4–2. Thus, a shift in the time domain corresponds to multiplication by the exponential ϵ^{-as} in the s-plane.

4.5.8 The Shift in the s-Plane

If a function is multiplied by ϵ^{-at}, the transform of the product is the transform of the original function with a added to every occurrence of s. We can use the definition of the transform to prove this as follows:

$$\mathcal{L}[\epsilon^{-at}f(t)] = \int_0^\infty \epsilon^{-at}f(t)\epsilon^{-st}\,dt$$

$$= \int_0^\infty f(t)\epsilon^{-(s+a)t}\,dt = F(s+a) \tag{4–27}$$

This relation is expressed as property G in Table 4–2. Thus, a shift in the complex frequency plane corresponds to multiplication by an exponential in the time domain.

We can use the s-plane shift transform property to develop other transforms. The following examples are illustrative.

EXAMPLE 4–13 The expression $\epsilon^{-at}\cos\omega t$ is often found in the response of electric circuits. What is the transform of this expression?

Given that we know the transform of $\cos\omega t$ (transform number 3), we can use the s-plane shift property (property G) to find the transform of the damped cosine function.

The transform of $\cos\omega t$ is

$$\mathcal{L}[\cos\omega t] = \frac{s}{s^2 + \omega^2}$$

Applying property G, we get

$$\mathcal{L}[\epsilon^{-at}\cos\omega t] = \frac{s+a}{(s+a)^2 + \omega^2} \tag{4–28}$$

This transform is number 7 in Table 4–1.

EXAMPLE 4–14 What is the transform of $\epsilon^{-at}\sin\omega t$?

This function, either by itself or, more commonly, with $\epsilon^{-at}\cos\omega t$, is one of the most likely transient responses of electric circuits. Using property G of Table 4–2 and transform pair 4 of Table 4–1, we obtain

$$\mathcal{L}[\epsilon^{-at}\sin\omega t] = \frac{\omega}{(s+a)^2 + \omega^2} \tag{4–29}$$

This relationship is transform pair 8 of Table 4–1.

EXAMPLE 4–15 Another possible factor in the transient response of circuits is $t\epsilon^{-at}$. What is its transform?

Using property G and transform pair 5, we have

$$\mathcal{L}[t\epsilon^{-at}] = \frac{1}{(s + a)^2} \tag{4–30}$$

This relationship is transform pair 6 of Table 4–1.

EXAMPLE 4–16 We have derived transform pairs for both $\cos \omega t$ and $\sin \omega t$. A more common excitation is a sinusoid with a phase shift. How do we find the Laplace transform of this function?

First we must find the function in terms of a sum of unshifted sine and cosine functions. If the excitation is

$$v(t) = 1.20 \sin(20t + 30°)$$

then the sine and cosine form is, using Equation (1–28),

$$v(t) = 1.039 \sin 20t + 0.600 \cos 20t$$

Now, using property A of Table 4–2 and transform pairs 3 and 4 of Table 4–1, we get

$$V(s) = \frac{20.78}{s^2 + 400} + \frac{0.600s}{s^2 + 400}$$

4.6 TRANSFORMATION OF DIFFERENTIAL EQUATIONS

Using the Laplace transform pairs and properties given in Tables 4–1 and 4–2, we can readily transform the integro-differential equations of electric circuits and solve for the transform of the variable. This is illustrated in several examples.

EXAMPLE 4–17 Transform the following equation, solve for $V(s)$, and then, using Tables 4–1 and 4–2, find $v(t)$:

$$\frac{dv}{dt} + 2v = 0, \qquad v(0^+) = 1$$

Properties A and C of Table 4–2 are used to get

$$sV(s) - 1 + 2V(s) = 0$$

Solving for $V(s)$, we obtain

$$V(s) = \frac{1}{s + 2}$$

Inspecting the transforms in Table 4–1, we see that pair 2 can be used to perform the inverse transformation to get

$$v(t) = \epsilon^{-2t}$$

This is the solution of the differential equation given. We were able to identify the proper inverse transform directly from the table to find the solution in the time domain. In most cases, however, we will find that the s-plane expression is more complicated. As a result, more algebraic manipulation may be needed to find expressions that appear in the table.

EXAMPLE 4–18

Transform the following equation and solve for $I(s)$:

$$40\frac{di}{dt} + 100i = 4, \qquad i(0^+) = 4 \text{ A}$$

We can use properties A and C of Table 4–2, transform pair 1 of Table 4–1, and the initial conditions to get

$$40[sI(s) - 4] + 100I(s) = \frac{4}{s}$$

Solving for $I(s)$, we obtain

$$I(s) = \frac{4(s + 0.025)}{s(s + 2.5)}$$

There is no s-plane expression in Table 4–1 that can be used directly to find the inverse transform of this polynomial fraction. An algebraic method of solution for expressions like this will be given in the next section.

EXAMPLE 4–19

Find $V(s)$ for the following differential equation and initial conditions:

$$3\frac{d^2v}{dt^2} + 2v = \sin 4t, \qquad v(0^+) = 0 \text{ V}, \qquad v'(0^+) = 1 \text{ V/s}$$

We can use properties A and D of Table 4–2, transform pair 4 of Table 4–1, and the initial conditions to get

$$3s^2V(s) - 3sv(0^+) - 3v'(0^+) + 2V(s) = \frac{4}{s^2 + 16}$$

Substituting the values of the initial conditions and solving for $V(s)$, the transform of the voltage, we obtain

$$V(s) = \frac{3s^2 + 52}{(s^2 + 16)(3s^2 + 2)}$$

Again, we cannot find an expression like this in Table 4–1.

4.7 INVERSION OF MORE COMPLICATED TIME FUNCTIONS

The general form of the transform of an equation with an unknown variable in the s-plane is

$$F(s) = \frac{a_0 s^n + a_1 s^{n-1} + a_2 s^{n-2} + \cdots + a_n}{b_0 s^m + b_1 s^{m-1} + b_2 s^{m-2} + \cdots + b_m} \tag{4–31}$$

This can be written more simply as

$$F(s) = \frac{N(s)}{D(s)} \tag{4–32}$$

where $N(s)$ and $D(s)$ are polynomials in s. When $F(s)$ is the response of a physical system to a real excitation, the degree of the numerator polynomial will always be equal to or less than that of the denominator polynomial.

The process of determining the time response from the s-plane response is called *inversion of the transform*. There are three methods for doing this. If the function is simple enough, it can be found along with its inverse, the time domain solution, in a table of Laplace transform pairs, as was done in Example 4–17. Or, the inverse transform may be found directly by using the definition of the inverse:

$$\mathcal{L}^{-1}[F(s)] = f(t) = \frac{1}{2\pi j} \int_{0 - j\infty}^{0 + j\infty} F(s)\epsilon^{st}\,ds$$

However, solution of this integral requires a knowledge of the mathematics of complex variables, which is beyond the scope of this text. When we must find the inverse transforms of more complicated expressions, we can use algebraic manipulation to find several simpler fractions that are more easily inverted.

4.7.1 Partial Fraction Expansion

The third method for finding the inverse of a transform, generally used in engineering applications, involves expanding the polynomial fraction $N(s)/D(s)$ so that the s-plane result is expressed as a sum of fractions whose numerators are constants and whose denominators are first-degree polynomials. This procedure is sometimes known as *Heaviside partial fraction expansion*. The inverse transforms of the expanded fractions are usually easily found using short tables of transform pairs and operations such as Tables 4–1 and 4–2. If the numerator of the unfactored fraction, $N(s)$, is of the same degree as the denominator, it must first be divided by the denominator to get a constant plus a proper fraction. When the equations are of real systems, the numerator will never be of higher degree, and will rarely be of the same degree, as the denominator.

EXAMPLE 4–20   Prepare the following fraction for Heaviside partial fraction expansion:

$$F(s) = \frac{10s^2 + 16s + 14}{s^2 - 4s - 5}$$

The degree of $N(s)$ is equal to that of $D(s)$, so we must divide the numerator by the denominator, which gives

$$F(s) = 10 + \frac{56s + 54}{s^2 - 4s - 5}$$

Using transform pair 9 of Table 4–1, we find that the inverse transform of the constant term, 10, is $10\delta(t)$. The next step is to factor the denominator polynomial of the fraction. Since the denominator is a quadratic polynomial of the form $s^2 + bs + c = 0$, we can use the quadratic formula given in Chapter 3, that is,

$$s_1, s_2 = -\frac{b}{2} \pm \sqrt{\left(\frac{b}{2}\right)^2 - c}$$

Methods for factoring higher order polynomials include using more complex formulas and calculator or computer programs. The programs usually give approximate results for the roots; when they are used properly, high accuracy is possible.

The roots of quadratic equations may assume three forms:

1. Real roots, none of which is repeated.
2. Real, repeated roots.
3. Complex or imaginary roots.

Complex or imaginary roots always occur in complex conjugate pairs. Complex roots could be repeated, but this is rare. The total number of roots will be equal to the degree of the polynomial and also to the number of independent energy storage components in the circuit or system.

The denominator must be factored to give a product of first-degree terms, such as

$$F(s) = \frac{N(s)}{A}(s + s_1)(s + s_2) \cdots (s + s_m) \tag{4–33}$$

where A is the coefficient of the highest degree term in the original unfactored denominator. The expanded function will be in the form

$$F(s) = \frac{K_1}{s + s_1} + \frac{K_2}{s + s_2} + \cdots + \frac{K_m}{s + s_m} \tag{4-34}$$

where $K_1, K_2, \ldots, K_m$ are constants called the *residues at the poles of* $F(s)$. The denominator terms $s + s_1, s + s_2, \ldots, s + s_m$ are the factors of the denominator in the original fraction. The preferred method of finding the numerator terms depends on the type of root involved.

The Laplace transform, $F(s)$, is a maximum at points where any of the denominator factors equal zero. In Equation (4-34), these points are where $s = -s_1, -s_2, \ldots, -s_m$. These are points on the complex frequency or s-plane where the response of the circuit has maxima. These points are called the poles of the response.

4.7.1.1   Real, Nonrepeated Roots   The procedure for expanding polynomial fractions with real, nonrepeated roots is best developed through an example.

EXAMPLE 4–21

If the factored polynomial fraction is

$$F(s) = \frac{6s + 10}{(s - 1)(s + 4)}$$

the expanded function will be in the form

$$F(s) = \frac{K_1}{s - 1} + \frac{K_2}{s + 4}$$

where K_1 and K_2 are constants. If we multiply both expressions for $F(s)$ by the factor $s - 1$ and equals the results, we get

$$\frac{(6s + 10)(s - 1)}{(s - 1)(s + 4)} = \frac{K_1(s - 1)}{s - 1} + \frac{K_2(s - 1)}{s + 4}$$

or

$$\frac{6s + 10}{s + 4} = K_1 + \frac{K_2(s - 1)}{s + 4}$$

Now if we evaluate this equation at $s = 1$, the location of the root due to the first factor in the denominator of the given factored polynomial fraction, the second term on the right drops out, and we can solve for K_1:

$$K_1 = [(s - 1)F(s)]_{s=1} = \left.\frac{6s + 10}{s + 4}\right|_{s=1} = \frac{16}{5}$$

Recapitulating, we have multiplied the s-domain response by the factor in the denominator of the fraction whose numerator we wish to evaluate. Then we have evaluated the resulting expression at the location of that root. We can find K_2 similarly:

$$K_2 = [(s + 4)F(s)]_{s = -4} = \frac{6s + 10}{s - 1}\bigg|_{s = -4} = \frac{14}{5}$$

We thus have, as the expanded fraction,

$$F(s) = \frac{16}{5(s - 1)} + \frac{14}{5(s + 4)}$$

That this is correct is easily verified by adding the two fractions:

$$F(s) = \frac{16(s + 4) + 14(s - 1)}{5(s - 1)(s + 4)} = \frac{6s + 10}{(s - 1)(s + 4)}$$

We can see that, for each root, this procedure eliminates the factor that causes $F(s)$ to become infinite at the s-plane location of the root. Consequently, we can now evaluate the numerator of each fraction separately. The operation that we have been performing may be expressed in the general formula

$$K_i = [(s + a)F(s)]_{s = -a} \tag{4–35}$$

which, when $F(s)$ is expanded, gives

$$F(s) = \frac{K_1}{s + s_1} + \frac{K_2}{s + s_2} + \cdots + \frac{K_i}{s + s_i}$$

EXAMPLE 4–22   Expand the following polynomial fraction, and find its inverse transform:

$$F(s) = \frac{7s + 5}{2(s + 4)(s + 3)(s + 2)}$$

The numerator is of lower order than the denominator, so we need not perform any preliminary division. Expanded, $F(s)$ will have the following form:

$$F(s) = \frac{K_1}{s + 4} + \frac{K_2}{s + 3} + \frac{K_3}{s + 2}$$

We can evaluate the constants directly, using Equation (4–35):

$$K_1 = \frac{7s + 5}{2(s + 3)(s + 2)}\bigg|_{s = -4} = \frac{-23}{4}$$

$$K_2 = \left.\frac{7s + 5}{2(s + 4)(s + 2)}\right|_{s = -3} = 8$$

$$K_3 = \left.\frac{7s + 5}{2(s + 4)(s + 3)}\right|_{s = -2} = \frac{-9}{4}$$

So

$$F(s) = -\frac{23}{4(s + 4)} + \frac{8}{s + 3} - \frac{9}{4(s + 2)}$$

Using transform pair 2 of Table 4–1 and operation pair A of Table 4–2, we can find the time domain equivalent of $F(s)$:

$$f(t) = -5.75\epsilon^{-4t} + 8\epsilon^{-3t} - 2.25\epsilon^{-2t}$$

This function is plotted in Figure 4–5.

4.7.1.2 Real, Repeated Roots Occasionally, real roots are repeated in the denominator of a polynomial fraction. An example that involves both repeated and non-repeated roots is

$$F(s) = \frac{N(s)}{(s + a)^n(s + b)(s + c)}$$

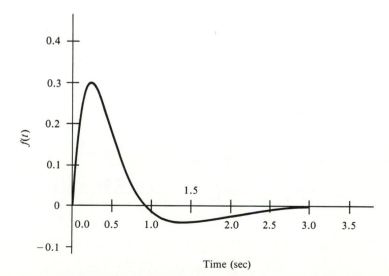

FIGURE 4-5 Plot of $f(t) = -5.75\epsilon^{-4t} + 8\epsilon^{-3t} - 2.25\epsilon^{-2t}$

If we try to apply the formula used for nonrepeated roots, we shortly see that we cannot eliminate all $(s + a)^n$ terms by multiplying once by $s + a$. As a result, the term will become infinite at $t = a$, and it will be impossible to eliminate one of the repeated roots. Accordingly, a different procedure is needed for repeated, real roots. For the foregoing equation for $F(s)$, the result will be in the following form:

$$F(s) = \frac{K_1}{(s + a)^n} + \frac{K_2}{(s + a)^{n-1}} + \cdots + \frac{K_n}{s + a} + \frac{K_{n+1}}{s + b} + \frac{K_{n+2}}{s + c}$$

Again, the numerators are constants.

The numerators of the terms due to the first-order poles can be found by using the method for real, nonrepeated roots, and those for the repeated roots may be found by using the formula

$$K_i = \frac{1}{(i - 1)!} \left| \frac{d^{i-1}(s + a)^n F(s)}{ds^{i-1}} \right|_{s = -a} \tag{4–36}$$

where i is the sequential number of the expanded fraction terms, n is the degree of the multiple root, and $-a$ is the location of the root in the s-plane. If $n = 1$, this reduces to Equation (4–35), the formula for nonrepeated terms. Equation (4–36) is thus a general formula for all cases.

EXAMPLE 4–23

If we need to expand

$$F(s) = \frac{4s^2 + 18s + 11}{(s + 2)^2(s + 1)}$$

the expanded terms will be in the form

$$F(s) = \frac{K_1}{(s + 2)^2} + \frac{K_2}{s + 2} + \frac{K_3}{s + 1}$$

where

$$K_1 = \frac{1}{0!} \cdot \frac{4s^2 + 18s + 11}{s + 1} \bigg|_{s = -2} = 9$$

$$K_2 = \frac{1}{1!} \cdot \frac{d}{ds} \left(\frac{4s^2 + 18s + 11}{s + 1} \right) \bigg|_{s = -2} = 7$$

and

$$K_3 = \frac{4s^2 + 18s + 11}{(s + 2)^2} \bigg|_{s = -1} = -3$$

Then

$$F(s) = \frac{9}{(s + 2)^2} + \frac{7}{s + 2} - \frac{3}{s + 1}$$

Using pairs 2 and 6 of Table 4–1 and operation A of Table 4–2, we get the time domain function

$$f(t) = 9t\epsilon^{-2t} + 7\epsilon^{-2t} - 3\epsilon^{-t}$$

This has three exponential terms, one of which is multiplied by t. The curve, which is plotted in Figure 4–6, decays asymptotically to zero as t approaches infinity. We have seen that decaying exponentials approach zero asymptotically as t approaches infinity. By L'Hospital's rule, $t\epsilon^{-at}$ will also decay asymptotically to zero as t increases.

4.7.1.3 Complex Roots   When the denominator of a factored polynomial fraction includes complex or imaginary roots, the procedure is basically the same as that used for real roots. Of course, the complex quantities involved make the evaluation more complicated; however, since complex roots occur in pairs, and since the numerators of each pair will be complex conjugates, only one of each pair needs to be evaluated.

If a system has only a single pair of complex roots, as

$$F(s) = \frac{N(s)}{(s + a + jb)(s + a - jb)}$$

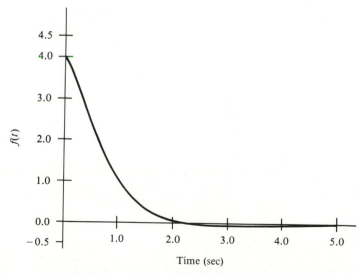

FIGURE 4–6   Plot of $f(t) = 9t\epsilon^{-2t} + 7\epsilon^{-2t} - 3\epsilon^{-t}$

then the expanded fraction will be in the form

$$F(s) = \frac{K_1}{s + a + jb} + \frac{K_2}{s + a - jb}$$

where K_1 and K_2 are complex. One method of dealing with this situation is that used with real, nonrepeated roots, but in the solution we must evaluate complex quantities. The numerators are

$$K_1 = [(s + a + jb)F(s)]_{s = -a - jb}$$

and

$$K_2 = [(s + a - jb)F(s)]_{s = -a + jb}$$

We can see by inspection that, for each complex factor that multiplies $F(s)$ in K_1, there is a corresponding complex conjugate factor in K_2, and the values entered to evaluate the residues K_1 and K_2 are also complex conjugates. As a result, K_2 equals K_1^*, the complex conjugate of K_1.

EXAMPLE 4–24 Expand

$$F(s) = \frac{s + 3}{(s^2 + 6s + 13)(s + 1)}$$

This equation has three roots, two for the second-degree term in the denominator and one for the first-degree term. Factoring the quadratic term, we get

$$s_1, s_2 = -\frac{6}{2} + \sqrt{\left(\frac{6}{2}\right)^2 - 13} = -3 \pm j2$$

The denominator terms are thus $s + 3 + j2$, $s + 3 - j2$, and $s + 1$, a pair of complex roots and a real root. The expanded fractions will then have the form

$$F(s) = \frac{K_1}{s + 3 + j2} + \frac{K_2}{s + 3 - j2} + \frac{K_3}{s + 1}$$

where

$$K_1 = \frac{s + 3}{(s + 3 - j2)(s + 1)}\bigg|_{s = -3 - j2} = 0.1768\angle 135.0°$$

$$K_2 = K_1^* = 0.1768\angle -135.0°$$

and

$$K_3 = \frac{s + 3}{s^2 + 6s + 13}\bigg|_{s = -1} = \frac{1}{4}$$

Then

$$F(s) = \frac{0.1768\angle 135.0°}{s + 3 + j2} + \frac{0.1768\angle -135.0°}{s + 3 - j2} + \frac{1}{4(s + 1)}$$

Using pair 2 of Table 4–1, we can get the time domain form

$$f(t) = 0.1768\angle 135°\epsilon^{-(3 + j2)t} + 0.1768\angle -135°\epsilon^{-(3 - j2)t} + 0.2500\epsilon^{-t}$$

Now, the phasors $0.1768\angle 135.0°$ and $0.1768\angle -135.0°$ can be expressed as $0.1768\epsilon^{j135.0}$ and $0.1768\epsilon^{-j135.0}$, respectively. So we obtain

$$f(t) = 0.1768\epsilon^{-3t}[\epsilon^{j(2t - 135.0°)} + \epsilon^{-j(2t - 135.0°)}] + 0.2500\epsilon^{-t}$$

Using an equation derived from Euler's relation, Equation (2–12), we get

$$f(t) = 0.3536\epsilon^{-3t}[\cos(2t - 135.0°)] + 0.2500\epsilon^{-t}$$

A plot of this function, the sum of an exponentially damped sinusoid and a decaying exponential, is given in Figure 4–7. The curve does not have any obvious sinusoidal character because of the long period of the sine wave in comparison with the exponential multiplier.

This method for dealing with complex roots uses the general procedure given for real roots and does not require a separate formula. However, it is tedious, and

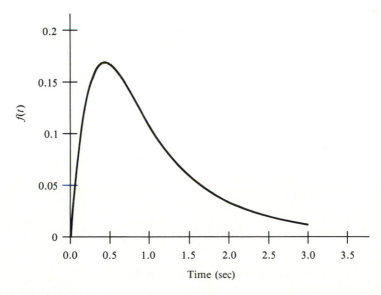

FIGURE 4–7   Plot of $f(t) = 0.3536\epsilon^{-3t}[\cos(2t - 135°)] + 0.2500\epsilon^{-t}$

its use is likely to introduce errors into the calculations. Another method that does involve another formula is developed in the appendix. The use of the added formula is simple, so the procedure is quicker and less error prone.

As a brief overview of this simpler method, if there is a pair of complex roots in the denominator, we can write the expression for the Laplace transform as

$$F(s) = \frac{Q(s)}{s^2 + bs + c} \tag{4-37}$$

Here, $Q(s)$ is the polynomial fraction that is left when $F(s)$ is multiplied by the factor that causes the complex roots. This factor, the denominator on the right-hand side of Equation (4-37), can be factored, giving

$$s^2 + bs + c = (s + a + j\omega)(s + a - j\omega)$$

Evaluation of $Q(s)$ at $s = -a + j\omega$ yields the complex number

$$Q(-a + j\omega) = M\angle\theta \tag{4-38}$$

Then the time domain factor due to the complex root is

$$f_1(t) = \frac{M}{\omega}\epsilon^{-at}\sin(\omega t + \theta) \tag{4-39}$$

$F(s)$ may include other terms whose inverse transforms will give the rest of the solution. So, using operation pair B of Table 4-2, we can obtain the inverse transform of $F(s)$ as the sum of the individual inverse transforms:

$$\mathcal{L}^{-1}[F(s)] = \mathcal{L}^{-1}[F_1(s)] + \mathcal{L}^{-1}[F_2(s)] = f_1(t) + f_2(t)$$

This method involves the use of an additional formula, but the formula is simple and much less computation is required.

EXAMPLE 4-25   Use the method just described to expand the s-plane expression given in Example 4-24.

We have

$$F(s) = \frac{s + 3}{(s^2 + 6s + 13)(s + 1)} = \frac{Q(s)}{s^2 + 6s + 13}$$

Factoring the denominator, we obtain

$$s^2 + 6s + 13 = (s + 3 + j2)(s + 3 - j2)$$

and therefore,

$$a = 3$$

and

$$\omega = 2$$

Factoring the quadratic term out of $F(s)$ yields

$$Q(s) = \frac{s + 3}{s + 1}$$

and

$$Q(-a + j\omega) = \frac{s + 3}{s + 1}\bigg|_{-3 + j2} = \frac{j2}{-2 + j2} = 0.707\angle -45°$$

So $M = 0.7071$ and $\theta = -45°$. Then, for the part of the solution due to the complex root pair, we have

$$f_1(t) = \frac{M}{\omega}\epsilon^{-at} \sin(\omega t + \theta) = \frac{0.7071}{2}\epsilon^{-3t} \sin(2t - 45°)$$

$$= 0.3536\epsilon^{-3t} \sin(2t - 45°)$$

For the rest of the response, namely, that due to the factor $s + 1$,

$$K_3 = \frac{s + 3}{s^2 + 6s + 13}\bigg|_{s = -1} = \frac{1}{4}$$

So

$$F_2 = \frac{1}{4(s + 1)}$$

Using transform pair 2 of Table 4–1, we get

$$f_2(t) = 0.2500\epsilon^{-t}$$

The complete solution is the sum of $f_1(t)$ and $f_2(t)$:

$$f(t) = 0.3536\epsilon^{-3t} \sin(2t - 45°) + 0.2500\epsilon^{-t}$$

This is equal to the result found in Example 4–24.

As can readily be seen, this method of solution is simpler than the one given earlier. Here again, the period of the sine wave is so long compared to the exponential multiplier that little evidence of it can be seen in the graph of the function.

4.8 THE INITIAL VALUE AND FINAL VALUE THEOREMS

Sometimes, it is only necessary to find the initial or the final value of a response. For a system that consists of a known electric circuit, the methods described in Section 3.7 can be used to find either value. If the time domain function for the response is known, substituting 0 or ∞ for t will give the initial and final response, respectively.

If the time domain function $f(t)$ for the response is not known, but its transform $F(s)$ is, the initial value theorem can be used to find the response at $t = 0$ seconds, without finding the inverse transform. The initial value theorem is

$$f(0^+) = \lim_{t \to 0^+} f(t) = \lim_{s \to \infty} sF(s) \qquad (4\text{-}40)$$

if the limit exists. As an example, let us find the initial value of a moderately complex response.

EXAMPLE 4–26   What is the initial value of the time domain expression if

$$F(s) = \frac{5s + 10}{s^2 + 8s + 15}$$

Applying the initial value theorem, we obtain

$$f(0^+) = \lim_{s \to 0} s\left|\frac{5s + 10}{s^2 + 8s + 15}\right| = \lim_{s \to \infty} \frac{5s^2 + 10s}{s^2 + 8s + 15}$$

Using L'Hospital's rule, we get

$$f(0^+) = 5$$

We can check this result by finding the inverse transform $f(t)$ and evaluating it at $t = 0$. Factoring $F(s)$, we have

$$F(s) = \frac{5(s + 2)}{(s + 3)(s + 5)}$$

Using partial fraction expansion, we obtain

$$F(s) = \frac{7.5}{s + 5} - \frac{2.5}{s + 3}$$

Laplace transform pair 2 of Table 4–1 gives the time domain response:

$$f(t) = 7.5\epsilon^{-5t} - 2.5\epsilon^{-3t}$$

Substituting 0 for t in this equation, we get

$$f(0^+) = 7.5 - 2.5 = 5$$

which is the same value that we found using the initial value theorem.

The final value theorem is

$$f(\infty) = \lim_{t \to \infty} f(t) = \lim_{s \to 0} sF(s) \qquad (4\text{–}41)$$

The final value theorem can be used as long as all poles of $sF(s)$ are in the left half-plane. It is often used in evaluating steady-state error in automatic control systems. As an example of the use of the theorem, we can find the final value of the output for a simple system.

EXAMPLE 4–27   Find the final time domain value for a system whose Laplace transform is

$$F(s) = \frac{10}{s(s + 2)}$$

Using the final value theorem, we get

$$f(\infty) = \lim_{s \to 0} sF(s) = \frac{10}{s + 2} = \frac{10}{2} = 5$$

As with the initial value theorem, we can readily check this result by finding the inverse transform of $F(s)$ and letting $t = \infty$ seconds. Applying partial fraction expansion to $F(s)$, we have

$$F(s) = \frac{10}{s(s + 2)} = \frac{5}{s} + \frac{5}{s + 2}$$

Transform pairs 1 and 2 of Table 4–1 give the inverse transforms of each of these terms. Combining them, we obtain

$$f(t) = 5 - 5\epsilon^{-2t}$$

and at $t = \infty$,

$$f(\infty) = 5$$

This is the same result as that found with the final value theorem.

The reason for the requirement that all poles be in the left half-plane is that poles on the $j\omega$ axis or in the right half-plane do not provide a limiting value for the response. If there are poles on the imaginary axis, the response includes sinusoids. If there are poles in the right half-plane, the response will increase without limit. Any single pole at the origin will provide a DC component, which will have a steady final value, so it will not interfere with the existence of a limit as long as the excitation has no DC component. In the expression to be evaluated, multiplication of the response function $F(s)$ by s will eliminate a single pole at the origin.

For simple functions such as those treated in Examples 4–26 and 4–27, there is not much time saved using the initial and final value theorems, but these theorems will save considerable time and effort when more complex systems are evaluated, particularly if the complete time response is not needed.

PROBLEMS

Using Table 4–1, find the transforms of the following time functions. (Assume all functions are zero for $t < 0$ seconds.)

4–1. 5

4–2. $7\epsilon^{-18t}$

4–3. $15 \cos 35t$

4–4. $25 \sin 42\pi t$

4–5. $25 \sin 35t$

4–6. $37 \cos 25t$

4–7. $35t$

4–8. $180t\epsilon^{-45t}$

4–9. $37.5\epsilon^{-22t} \cos 45t$

4–10. $29\epsilon^{-35t} \sin 32t$

4–11. $32\delta(t)$

4–12. $85\delta(t)$

Using Tables 4–1 and 4–2, transform the following. (Assume all functions are zero for $t < 0$ seconds, unless otherwise specified.)

4–13. $8t + 5$

4–14. $9\delta(t) + 7 + 2t$

4–15. $7 - 8t$

4–16. $7U(t - 7) + 4t - 20$

4–17. $\int_0^t 5U(t)\, dt$

4–18. $45\dfrac{dU(t - a)}{dt}$, $a = 7$

4–19. $23\dfrac{d5(t-a)}{dt}$, $a = 5$

4–20. $38\epsilon^{-20t}\sin 45t$

4–21. $(45t - 45)U(t - 1)$

4–22. $22\epsilon^{-10t}\cos 35t$

4–23. Find the Laplace transform for the waveform shown in Figure 4P–1.

4–24. Find the transform for the current shown in Figure 4P–2.

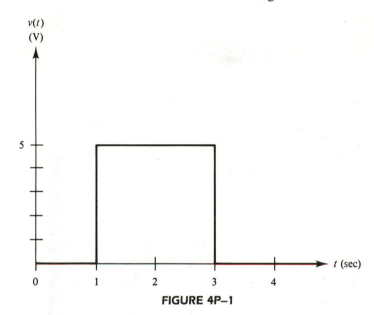

FIGURE 4P–1

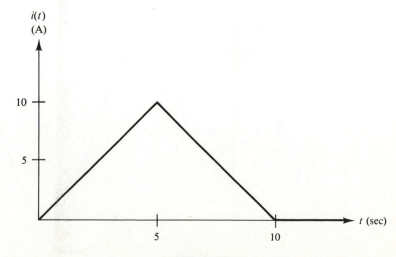

FIGURE 4P–2

4–25. Transform the voltage waveform shown in Figure 4P–3.

4–26. Transform the sinusoidal curve segment shown in Figure 4P–4.

Find the roots of the following polynomials:

4–27. $s^2 + 8s + 16$

4–28. $s^2 + 14s + 49$

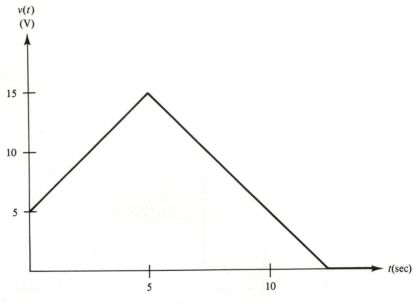

FIGURE 4P–3

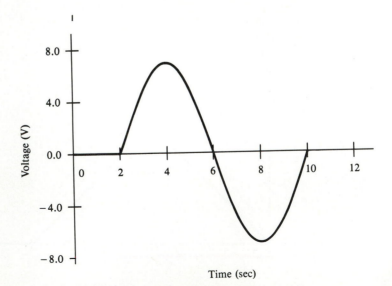

Time (sec)

FIGURE 4P–4

4–29. $s^2 + 9$

4–30. $s^2 + 4s + 25$

Expand the following polynomial fractions:

4–31. $F(s) = \dfrac{5s}{s^2 + 17s + 72}$

4–32. $F(s) = \dfrac{11s + 47}{s^2 + 8s + 7}$

4–33. $F(s) = \dfrac{5}{s^2 + 4s + 13}$

4–34. $F(s) = \dfrac{s + 3}{s^2 + 10s + 25}$

4–35. $F(s) = \dfrac{s + 3}{s^2 + 6s + 36}$

4–36. $F(s) = \dfrac{10(s + 7)}{(s + 1)(s^2 + 12s + 52)}$

Find the inverse transforms of the following expressions:

4–37. $F(s) = \dfrac{s^2 + 10}{s^2 + 8s + 12}$

4–38. $F(s) = \dfrac{10(s - 1)}{s^3 + 2s^2 + 5s}$

4–39. $F(s) = 10 - \dfrac{15}{s}$

4–40. $F(s) = \dfrac{17}{s + 20}$

4–41. $F(s) = 1/(35s + 1)$

4–42. $F(s) = \dfrac{15s + 50}{s^2 + 25}$

4–43. $F(s) = \dfrac{20}{s^3 + 7s^2 + 10s}$

4–44. $F(s) = \dfrac{2s + 4}{(s^2 + s)(s + 3)}$

4–45. $F(s) = \dfrac{2s^2 + 11s + 4}{s(s + 1)}$

4–46. $F(s) = \dfrac{49(s^2 + 10s)}{s^3 + 10s^2 + 25s}$

4–47. $F(s) = \dfrac{5}{s^2 + 36}(1 - \epsilon^{-\pi s/3})$

4–48. $F(s) = \dfrac{s + 3}{s^2 + 4s + 4}$

4-49. $F(s) = \dfrac{1}{(s + 2)^2(s^2 + 2s + 5)}$

4-50. $F(s) = \dfrac{5s + 40}{s^2 + 4s + 29}$

Find the initial and, if they exist, the final, values of the functions in:

4-51. The s-plane expressions in Problems 4–31, 4–33, and 4–35.

4-52. The expressions in Problems 4–32, 4–34, and 4–36.

CHAPTER 5

Circuit Solutions Using Laplace Transforms

5.1   OBJECTIVES

On completion of this chapter, you should be able to:

- Transform the components and sources of a circuit into the correct s-plane representation.
- Insert initial condition sources into the s-plane circuit.
- Solve for the transient response currents and voltages in the transformed circuit.
- Find the time domain solutions for the transient responses of circuits.
- Be familiar with the general relationship between circuit components and the transient responses of the circuit.

5.2   INTRODUCTION

In Chapter 3, the differential equations describing the responses of linear electrical networks and mechanical systems were developed. These differential equations are used to find the complete response for any system that has both energy storage components, such as inductors and capacitors, and energy-dissipating components, such as resistors. In simple cases, the equations may be solved directly, sometimes even from memory. For some complex systems, it is possible to find an approximate solution using data from general tables or curves, but this is impractical in many cases. Mathematically oriented engineers have usually handled these problems by writing the differential equations of the system and then transforming the equations directly to find the responses. For those who are more practically oriented, a method is available in which the circuit itself is transformed and standard circuit analysis techniques are used to find the responses.

5.3   CIRCUIT TRANSFORMATION

A circuit is transformed by replacing the time domain representations of impedances, sources, and initial conditions by their s-domain equivalents. In most cases, this can be done by means of Tables 4–1 and 4–2. Once transformation is accomplished, standard circuit analysis laws and techniques, such as Kirchhoff's voltage law and mesh analysis, may be applied to derive algebraic equations for the desired unknowns in the complex frequency domain. These equations can be readily solved for the desired unknowns.

The results will be in the s-plane, so an inverse transformation must be performed to find the time domain solutions. The complete procedure involves the use of basic circuit theory, tables of Laplace transform pairs, and algebraic and trigonometric operations. No differential equations need to be written. The transformed circuit is in a familiar form, and initial conditions are easily included in it.

5.3.1   *s*-plane Equivalent of Sources

Both current and voltage sources are transformed in a similar manner. The letters used to represent excitation sources in the transformed circuits will be those used for continuous sources in steady-state circuit diagrams. To differentiate them from the time domain circuit symbols, uppercase letters are used.

The transformed sources will often have zero output until they are turned on at some specified time. The complete time domain expression for such a switched excitation includes a step function multiplier to turn a continuous function on.

In tables of transform pairs, the step function is often not written in the time domain expressions. Usually, the table includes a statement that all time functions are zero for $t < 0$ seconds, and in printing the tables, the multiplier $U(t)$ is not included. For instance, in Table 4–1, the inverse transform of $1/s$ is said to be 1, but $1/s$ is actually the transform of the unit step function.

This convention of leaving out the step function usually causes no problems when finding the Laplace transform. It may cause errors, however, when inverting the transforms to get the time domain solution, unless the fact that the responses are all zero for $t < 0$ seconds is kept in mind. As a rule, in solving transient problems, the unit step function is not written when describing the excitation in the equations, unless it occurs at some time other than $t = 0$ seconds.

5.3.1.1   DC Source Switched On at $t = 0$ Seconds   The simplest excitation source found in transient problems is a DC source turned on at $t = 0$ seconds. The following pair of time domain expressions can be used to describe this type of source:

$$e(t) = A, \qquad \text{for } t > 0 \text{ seconds}$$

$$e(t) = 0, \qquad \text{for } t < 0 \text{ seconds}$$

These may be written more simply as

$$e(t) = AU(t)$$

where $U(t)$ is the unit step function. The effect of the step function is to switch the DC source on at $t = 0$ seconds. The source can be produced by the circuit shown in Figure 5–1(a), with the step function represented by the switch. As mentioned earlier, in writing the equations, the step function multiplier is usually omitted.

Using transform pair 1 from Table 4–1 and operation A from Table 4–2, we obtain

$$E(s) = \mathcal{L}[A] = \frac{A}{s} \tag{5–1}$$

This is the s-plane or Laplace transform representation of $AU(t)$; it is shown in Figure 5–1(b).

5.3.1.2 Sinusoidal AC Source Switched On at $t = 0$ Seconds Another very common type of source is the switched sinusoidal wave source, shown in Figure 5–2(a). The time domain expression for this source is

$$e(t) = A\ \sin(\omega t + \theta)$$

In Table 4–1, there is no transform pair for a phase-shifted sinusoid, but, using a well-known trigonometric identity and Equation (1–28), we can express this sinusoid as

$$e(t) = B\ \cos \omega t + C\ \sin \omega t \tag{5–2}$$

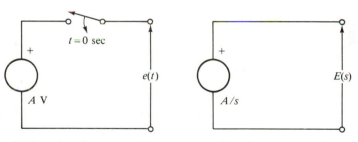

(a) Time domain representation   (b) s-plane representation

FIGURE 5–1 DC source switched on at $t = 0$ seconds

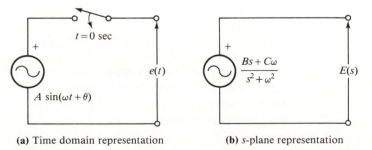

(a) Time domain representation   (b) s-plane representation

FIGURE 5–2 Sinusoidal AC source switched on at $t = 0$ seconds

where

$$B = A \sin \theta$$

and

$$C = A \cos \theta$$

If A and θ are known, B and C can readily be found using a scientific calculator. The two terms in Equation (5–2) can easily be transformed using transform pairs 3 and 4 of Table 4–1 and operation A of Table 4–2. Transforming gives

$$E(s) = \mathcal{L}[A \sin(\omega t + \theta)] = \frac{Bs + C\omega}{s^2 + \omega^2} \tag{5–3}$$

as the s-plane equivalent. This transform can be readily seen to be the sum of the transforms of $B \cos \omega t$ and $C \sin \omega t$. This is shown in Figure 5–2(b).

EXAMPLE 5–1   Find the Laplace transform of $e(t) = 25 \sin(60t + 35°)$.

Here, $A = 25$ and $\theta = 35°$. Entering these values into the calculator and performing a polar-to-rectangular conversion, we get, as the horizontal or σ component, 14.34. This is B, the coefficient of the cosine term in Equation (5–2). The vertical or $j\omega$ component is C, the coefficient of the sine term in Equation (5–2). This equals 20.48. Then, using the transform pair in Equation (5–3), we obtain

$$E(s) = \mathcal{L}[25 \sin(60t + 35°)] = \frac{14.34s + 20.48(60)}{s^2 + 60^2} = \frac{14.34(s + 85.69)}{s^2 + 3600}$$

5.3.1.3 Decaying Exponential Source Turned On at $t = 0$ Seconds   The switched exponential is a less common type of excitation than DC or sinusoidal AC that occurs when a switched DC source drives a simple high-pass RC network. A realistic example is a very long rectangular pulse that drives a capacitive coupling network, as shown in Figure 5–3(a). The network output during the time the pulse is on will be a decaying exponential, switched on at the time the input pulse starts. Figure 5–3(b) shows such a switched exponential source, whose time domain expression is

$$e(t) = A\epsilon^{-at}$$

If a is positive, the function approaches zero as time approaches infinity. As a result, there will be no steady-state currents or voltages in a circuit with this type of excitation. Using transform pair 2 of Table 4–1 and operation A of Table 4–2, we obtain the s-plane expression:

$$E(s) = \mathcal{L}[A\epsilon^{-at}] = \frac{A}{(s + a)} \tag{5–4}$$

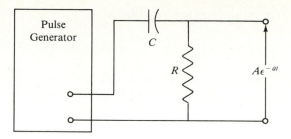

(a) Circuit diagram for switched exponential source

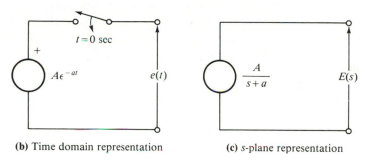

(b) Time domain representation   **(c)** s-plane representation

FIGURE 5–3   Decaying exponential source switched on at $t = 0$ seconds

Figure 5–3(c) shows the s-plane representation of an exponential excitation source. If $a = 0$, Equation (5–4) becomes

$$E(s) = \frac{A}{s} \qquad (5\text{–}5)$$

which is the s-plane representation of a step function. Thus, transform pair 1 of Table 4–1 is a special case of transform pair 2 of the same table.

5.3.1.4   Sources Switched On at $t = a$ Seconds   Occasionally, a source may be turned on at some time other than $t = 0$ seconds. If the source in question is the only switched source, time may be redefined so that switching occurs at $t = 0$ seconds, and then the transform pairs in Table 4–1 may be used. If two sources are switched on at different times, the time shift operation F of Table 4–2 can be used in transforming one or both of the sources. In this case, it is desirable to use the step function multiplier when expressing the source in the time domain. For example, the first of two switched DC sources that are added together to produce the source $e(t)$ is

$$e_1(t) = AU(t)$$

and the second is

$$e_2(t) = BU(t - a)$$

The complete time function is

$$e_3(t) = AU(t) + BU(t - a)$$

and is illustrated in Figure 5–4. Transforming this function by means of pair 1 of Table 4–1 and operations A and F of Table 4–2, we get

$$E_3(s) = \frac{A}{s} + \frac{B\epsilon^{-as}}{s}$$

If the functions that are switched vary with time, the situation is more complicated. For instance, if we have a voltage that is the product of a step function $AU(t)$ occurring at 0 seconds and a sinusoid $BU(t - a) \cos(\omega t + \theta)$ switched on at a seconds, we can describe these sources by

$$e_1(t) = AU(t)$$

and

$$e_2(t) = BU(t - a) \cos(\omega t + \theta)$$

where θ is the angle of the cosine wave at $t = 0$ seconds. To use the shifting operation pair F of Table 4–2, a factor equal to the time shift a must be subtracted from t wherever it occurs in the expression for the components of the unshifted wave. So we need to write

$$e_2(t) = BU(t - a) \cos[\omega(t - a) + \phi]$$

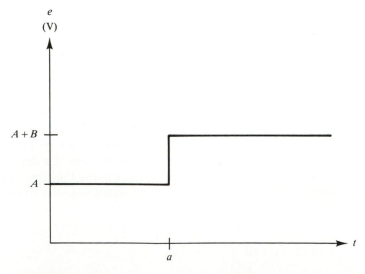

FIGURE 5–4 Sum of a step function and a delayed step function

or

$$e_2(t) = BU(t - a) \cos(\omega t + \phi - \omega a)$$

where

$$\phi - \omega a = \theta$$

Then we need to replace the phase-shifted sinusoid by the sum of a sine and a cosine function, as we did in Section 5.3.1.2. When we do, we get

$$e_2(t) = [C \cos(\omega t + \phi - \omega a) + D \sin(\omega t + \phi - \omega a)]U(t - a)$$

where

$$C = B \sin(\phi - \omega a)$$

and

$$D = B \cos(\phi - \omega a)$$

Using pairs 3 and 4 and operations A and F to transform the above, we have, for the transform of E_2,

$$E_2(s) = \frac{B[s \sin(\phi - \omega a) + \omega \cos(\phi - \omega a)]\epsilon^{-(\phi - \omega a)s}}{s^2 + \omega^2}$$

Note that when several excitation sources in a system are turned on at different times, we need to use the shifted step function in representing one of the sources.

5.3.2 *s*-plane Equivalent of Impedance

We can find impedances in the s-plane directly by first writing the general equation for the relationship between voltage and current for the individual circuit elements. Then we produce the s-plane equivalents using Laplace transform pairs and operations. Finally, we use Ohm's law to solve for the impedances in the s-domain:

$$Z(s) = \frac{V(s)}{I(s)}$$

In this form of the law, each time domain expression is replaced by its s-domain equivalent.

5.3.2.1 Resistance Ohm's law applied to the circuit shown in Figure 5–5(a) yields, for the relationship between voltage across and current through the resistance,

$$v(t) = i(t)R$$

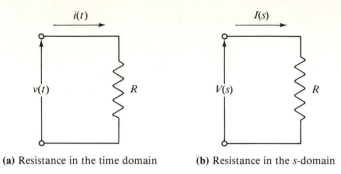

(a) Resistance in the time domain   **(b)** Resistance in the *s*-domain

FIGURE 5–5 Transformation of resistance

Transforming this to the *s*-plane, we get

$$V(s) = I(s)R$$

Here, the current and voltage are directly proportional to each other. That is, if one is switched on at $t = a$ seconds, the other is also; and if one is a sinusoid with a phase shift of $\theta°$, the other is also. From Ohm's law, the *s*-plane impedance is the voltage divided by the current:

$$Z(s) = R(s) = \frac{V(s)}{I(s)} = R \tag{5–6}$$

The *s*-plane representation of resistance is shown in Figure 5–5(b).

5.3.2.2 Capacitance The relationship between voltage and current in an initially uncharged capacitor, as shown in Figure 5–6(a), is given by Equation (1–37):

$$v(t) = \frac{1}{C}\int_0^t i\, dt$$

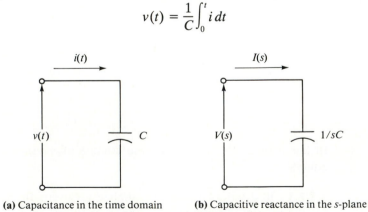

(a) Capacitance in the time domain   **(b)** Capacitive reactance in the *s*-plane

FIGURE 5–6 Transformation of capacitance

Transforming this equation by using operations A and E, we get

$$V(s) = \frac{1}{C} \cdot \frac{I(s)}{s} = \frac{I(s)}{sC}$$

This can be solved for the s-plane impedance:

$$Z(s) = \frac{V(s)}{I(s)} = \frac{1}{sC} \tag{5-7}$$

The transformed capacitance is shown in Figure 5–6(b). The complex frequency s equals $\sigma + j\omega$, so if we let $\sigma = 0$, s will be $j\omega$. We have thus found the impedance to a sine wave whose frequency in radians is ω. Then the s-domain impedance is

$$Z(j\omega) = X_{\mathrm{L}}(j\omega) = \frac{1}{j\omega C}$$

which is the impedance of a capacitance to a sinusoid.

5.3.2.3 Inductance Figure 5–7(a) shows an initially unenergized inductance. The voltage developed across the inductance due to a current flowing through it is found using Equation (1–50):

$$v(t) = L\frac{di}{dt}$$

Transforming this equation using operation pairs A and C, we get

$$V(s) = sLI(s)$$

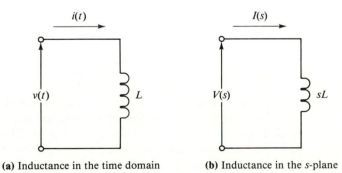

(a) Inductance in the time domain   (b) Inductance in the s-plane

FIGURE 5–7 Transformation of inductance

In an initially unenergized inductor, $i(0^+)$ is zero, so the second term in the s-plane form for pair C, $f(0^+)$, can be omitted. Using Ohm's law, we obtain, for the s-plane impedance,

$$Z(s) = \frac{V(s)}{I(s)} = sL \tag{5–8}$$

This is the s-domain equivalent of inductance and is illustrated in Figure 5–7(b). If $\sigma = 0$, then $s = j\omega$, and the s-plane impedance is

$$Z(j\omega) = X_L(j\omega) = j\omega L$$

which is the AC sinusoidal impedance, as expected.

5.3.3   Initial Conditions

Reactive components can store energy, the presence of which can be detected by sensing the voltages across capacitors or the currents through inductors. We must account for these voltages and currents when we apply Kirchhoff's voltage or current law to find the transient response of networks, since they represent energy stored in the components. To do this in the s-plane, we must find the s-domain equivalents of these voltages and currents.

5.3.3.1   Initially Charged Capacitors   If the capacitor in Figure 5–8(a) has an initial charge, then the expression for the total voltage will be

$$v(t) = \frac{1}{C}\int_0^t i\,dt + v(0^+) \tag{5–9}$$

where $v(0^+)$ is the initial voltage at $t = 0^+$ seconds. The term $1/C \int i\,dt$ is the change in voltage resulting from current flow after $t = 0$ seconds. From Kirchhoff's voltage law, the two terms may be considered separately, as illustrated in Figure 5–8(b). Transforming the expression for the total voltage using pair 1 and operation K, we get

$$V(s) = \frac{I(s)}{sC} + \frac{v(0^+)}{s} \tag{5–10}$$

This sum of voltages resembles an expression of Kirchhoff's voltage law in the s-domain. The first term is the expression for the voltage across an initially uncharged capacitor, the second the voltage due to the initial charge on the capacitor. These two terms may be represented by a series circuit, as shown in Figures 5–8(b) and 5–8(c). This form of circuit may easily be used directly in series applications, such as in mesh analysis.

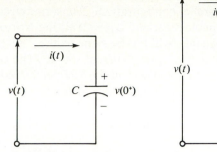

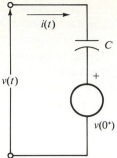

(a) Initially charged capacitor

(b) Thevenin equivalent of
initially charged capacitor
in time domain

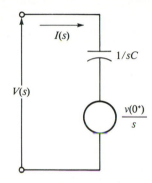

(c) Thevenin equivalent of initially
charged capacitor in s-plane

FIGURE 5–8 *Transformation of initially charged capacitor*

EXAMPLE 5–2 Let the circuit in Figure 5–8(a) show a capacitor with an initial voltage $v(0^-) = v(0^+)$ and an input current $i(t)$. Give the equations for the capacitor voltage in the time domain and in the s-plane, if $C = 10$ μfarads and $v(0^+) = 5$ volts.

Using Equation (5–9), we get the following equation for the voltage in the time domain:

$$v(t) = \frac{1}{C}\int_0^t i\, dt + v(0^+) = \frac{1}{10^{-5}}\int_0^t i\, dt + 5$$

In the s-plane, we can use Equation (5–10) to get

$$V(s) = \frac{I(s)}{10^{-5}s} + \frac{5}{s}$$

If a parallel circuit is to be analyzed, it is simpler to use an equivalent circuit that has a parallel configuration. The series circuit of Figure 5–8(c) is a Thévenin circuit. Its simplest parallel equivalent, the Norton equivalent, is easily found. The Norton impedance will be the same as the Thévenin impedance, $1/sC$, and will be in parallel with the Norton current source. The current source is equal to the short circuit current that will flow in the Thévenin circuit at $t = 0$ seconds. The circuit for this condition is shown in Figure 5–9.

The short circuit current can be seen to be the initial charge voltage $v(0^+)/s$ divided by the s-plane impedance $1/sC$. This gives

$$I_{SC} = I_N = \frac{v(0^+)/s}{1/sC} = Cv(0^+) \tag{5–11}$$

In the Norton equivalent circuit, this current is in parallel with the uncharged capacitance, as shown in Figure 5–10.

Using transform pair 9, we can find the time domain equivalent of the short circuit current:

$$i(t) = Cv(0^+)\delta(t) \tag{5–12}$$

The current source, an impulse function, is illustrated in Figure 5–11.

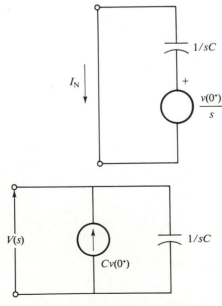

FIGURE 5–9 Finding the Norton equivalent for initially charged capacitance

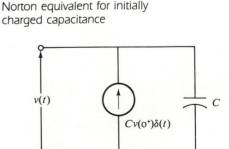

FIGURE 5–10 Norton equivalent of initially charged capacitance in the *s*-plane

FIGURE 5–11 Norton equivalent of initially charged capacitance in the time domain

EXAMPLE 5–3 Find the parallel equivalent of the circuit in Example 5–2. Express it in the time domain and the s-plane.

In the time domain, we use Equation (5–12):

$$i(t) = Cv(0^+)\delta(t) = 10 \times 10^{-6}(5)\delta(t) = 5 \times 10^{-5}\delta(t)$$

To get the s-plane expression, we use Equation (5–11):

$$I(s) = Cv(0^+) = 10 \times 10^{-6}(5) = 5 \times 10^{-5}$$

The parallel equivalent is simpler than the series equivalent found in Example 5–2.

5.3.3.2 Initial Current Through an Energized Inductor The inductor shown in Figure 5–12(a) is initially energized by means of a current $i(0^+)$ at $t = 0^+$ flowing through it at $t = 0^+$ seconds. When an external voltage is applied, a transient current builds up through the inductor until the forced current is flowing. The current actually flowing at any time is the sum of the initial and the transient components. The total current will be

$$i(t) = \frac{1}{L}\int_0^t v\,dt + i(0^+) \tag{5–13}$$

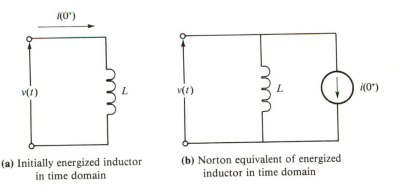

(a) Initially energized inductor in time domain

(b) Norton equivalent of energized inductor in time domain

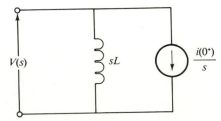

(c) Norton equivalent of energized inductor in s-plane

FIGURE 5–12 Finding the Norton equivalent of initially energized inductor

where $i(0^+)$ is the initial current. The term $1/L \int v \, dt$ accounts for the change in current after $t = 0$ seconds.

Kirchhoff's current law indicates that the sum given in Equation (5–13) can be represented by the initial source $i(0^+)$ in parallel with an unenergized inductor, as shown in Figure 5–12(b). We can transform the time domain function using transform pair 1 and operation E. We get

$$I(s) = \frac{V(s)}{sL} + \frac{i(0^+)}{s} \tag{5–14}$$

The circuit that this equation depicts is illustrated in Figure 5–12(c). The term $i(0^+)/s$ represents a step function in the time domain. Since this is a parallel circuit, it is the easiest configuration to use in nodal analysis.

EXAMPLE 5–4   The inductor in Figure 5–12(c) is 7 henrys, and the initial current is 5 amperes. Express the total current in the time domain and the s-plane.

The time domain current is found using Equation (5–13):

$$i(t) = \frac{1}{7} \int_0^t v \, dt + 5$$

To get the s-plane equivalent, we use Equation (5–15):

$$I(s) = \frac{V(s)}{7s} + \frac{5}{s}$$

We could have used transform pair 1 and operation E on the time domain equation to get the same answer.

The parallel equivalent circuit is a Norton circuit, and its series equivalent is a Thévenin circuit. The latter consists of the impedance sL in series with a voltage source whose amplitude is the open circuit voltage of the Norton circuit, as shown in Figure 5–13. This voltage is

$$V_{\text{Th}}(s) = V_{\text{OC}} = \frac{i(0^+)}{s} \cdot sL = Li(0^+) \tag{5–15}$$

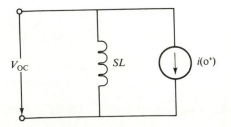

FIGURE 5–13 Finding the Thévenin voltage for an energized inductor

Hence, the Thévenin equivalent of an initially energized inductor in the s-domain is an unenergized inductor in series with an impulse voltage source, as shown in Figure 5–14.

From transform pair 9 and operation A, in the time domain the Thévenin voltage is

$$v_{Th}(t) = Li(0^+)\delta(t) \qquad (5\text{--}16)$$

as shown in Figure 5–15. Here again, we have the impulse function used in the representation of an initial condition. The integral of the impulse function is its area:

$$\int_0^t Li(0^+)\delta(t)dt = Li(0^+)U(t) \qquad (5\text{--}17)$$

This is a step function of amplitude $Li(0^+)$. We have used the unit step function in the expression to emphasize the fact that the initial condition source is turned on at $t = 0$ seconds. Note that it is physically impossible to have a step function of current through an inductor, since that would require an instantaneous change in energy, which in turn would require infinite power. Therefore, it is no surprise that we have to use a physically impossible function, the impulse function, to represent the current through an inductor.

EXAMPLE 5–5   Find the series equivalent for the initially energized inductance in Example 5–4, shown in Figure 5–12(c).

From Equation (5–17), we have, in the time domain,

$$v_{Th}(t) = Li(0^+)\delta(t) = 7(5)\delta(t)$$

and Equation (5–16) gives us,

$$V_{Th}(s) = Li(0^+) = 35$$

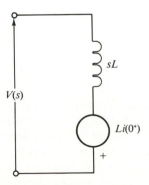

FIGURE 5–14 Thévenin equivalent of energized inductor in s-plane

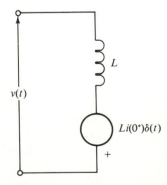

FIGURE 5–15 Thévenin equivalent of energized inductor in time domain

Clearly, in the s-plane, the series equivalent for the energized inductor is simpler than the parallel equivalent.

5.3.4 Initial and Final Circuits

The initial circuit simplifies the process of finding initial conditions in circuits that include energy storage elements. The initial and the final circuit may be useful in finding the states of the circuit both before the start of a transient and after it has reached its final state. The inductors and capacitors may be replaced by open or short circuits, which simplifies analysis of the circuits.

5.3.4.1 The Initial Circuit In Chapter 1, the presence of charges on the plates was said to provide the opposition to current flow in capacitors. When the capacitor is uncharged there is no opposition to current. This is because it has no voltage across it until the charge starts to build up. When uncharged, it can be replaced by any element that has similar characteristics, i.e., a short circuit. In the same chapter, it was also said that the current through an inductor cannot be changed instantaneously. Thus, when the initial current is zero, the inductor can be replaced by an element no current can flow through, i.e., an open circuit. Figure 5–16(a) shows a circuit with an initially uncharged capacitor and an inductor with no initial current flow. The initial circuit for this circuit is shown in Figure 5–16(b). The switch has just been closed and no longer opposes the flow of current, so it can be replaced by a short circuit. The inductor has had no current flow up to the time of closure of the switch, so it can be replaced by an open circuit. The uncharged capacitor can be replaced by a short circuit. The inductor will prevent current flow through itself and R_3, so all the current flows through R_1 and R_2 in series. The source voltage is divided across those resistors. When a capacitor or inductor has an initial condition voltage or current, for the first instant of the transient period, the initial condition can be replaced by the initial condition circuits shown in Figures 5–8 through 5–15 with the uncharged capacitors and unfluxed inductors replaced by short circuits and open circuits, respectively.

5.3.4.2 The Final Circuit When fully charged, a capacitor has no current flowing into it, so it can be replaced by an open circuit. When an inductor is fully energized, its current is constant, so it provides no opposition to the flow of current. It can thus be replaced by a short circuit without affecting the rest of the circuit in any way. Applied to the circuit shown in Figure 5–16(a), these facts give the equivalent circuit shown in Figure 5–16(c). Now, all the current flows through R_1 and R_3 in series. This circuit can also be called the *steady-state circuit*. As a rule, when the excitation is a sinusoid or some other continuous time-varying function, this technique will not give the proper results. For sinusoidal excitation, steady-state AC circuit analysis should be used.

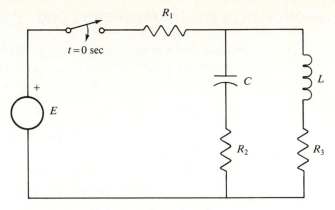

(a) RLC circuit

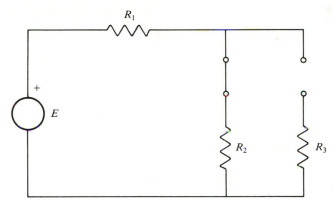

(b) Initial circuit

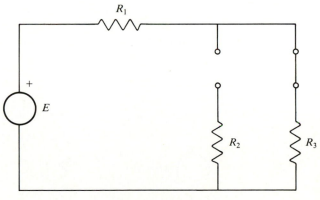

(c) Final circuit

FIGURE 5–16   Initial and final circuits

5.4 TRANSFORMATION OF COMPLETE CIRCUITS

Now that we have a way of expressing sources, components, and initial conditions in the s-plane, we are ready to transform complete circuits. The procedure for transforming any time domain circuit to the s-domain is as follows:

1. Study the problem to determine the best approach to the desired solution.
2. Redraw the circuit with the initial condition sources in the desired form, which will depend on the method of solution chosen in step 1.
3. Label each component with its s-plane equivalent.
4. Label each source with its s-domain equivalent.
5. Calculate the s-plane magnitude of the initial condition sources, and insert them in the drawing.
6. Use the method of solution chosen in step 1 to develop the equation or equations of the system. These will be algebraic equations in s that can be solved for the desired unknown. The solution will be expressed in the s-domain. We still, however, will need the time domain solution to get a result we can understand. If the s-plane function is simple, we can find it in the table of Laplace transform pairs. Then, the time domain solution is easily found. More complete tables of transform pairs are available, but no such table can be exhaustive, and large tables with many pairs are difficult to use. When the proper pair is not available, further algebraic and, sometimes, trigonometric manipulation is required.

The method used to find the s-plane solution is illustrated in the following example.

EXAMPLE 5–6 Find the s-plane current through the inductance in the circuit shown in Figure 5–17. The switch is closed at $t = 0$ seconds.

The circuit that produces the initial condition is shown in Figure 5–18. If we use the Thévenin equivalent for $i_L(0^+)$, the equivalent circuit used to solve for the

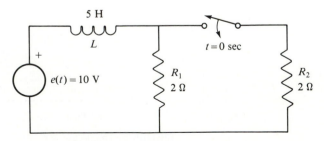

FIGURE 5–17 Circuit for Example 5–6

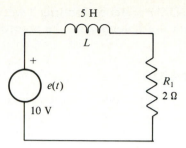

FIGURE 5–18 Equivalent circuit at $t = 0^-$ sec

transient solution will be a simple series circuit. This is shown in Figure 5–19. Here, the initial condition source is, from Equation (5–16),

$$v(0^+) = Li(0^+)\delta(t) = 25\delta(t)$$

Note that R_t in the figure is R_1 in parallel with R_2, or 1 Ω.

With this simple series circuit, the solution for the current $I(s)$ can be found using Ohm's law. With the s-plane circuit shown in Figure 5–20, we have

$$I(s) = \frac{E(s)}{Z(s)} = \frac{10/s + 25}{5s + 1} = \frac{5(s + 0.4)}{s(s + 0.2)}$$

where $E(s)$ includes both the excitation, $10/s$, and the initial condition source, 25, in series. This is a moderately complex s-plane function that is not in Table 4–1, our

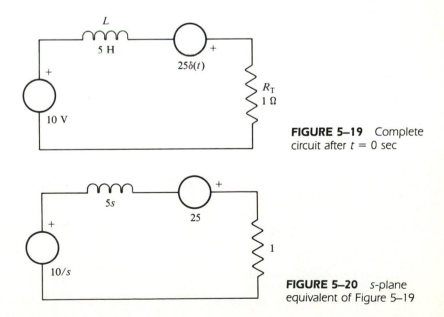

FIGURE 5–19 Complete circuit after $t = 0$ sec

FIGURE 5–20 s-plane equivalent of Figure 5–19

table of transform pairs. The method of solving it will be shown in the next section. As long as we can factor the denominator, the procedure will work for any *s*-plane expression.

5.4.1 Solution of First-Order Circuits

We learned in Chapter 3 how to solve first-order circuits with step functions applied. These circuits are simple enough to be solved by memory, since the solutions consist of simple exponential curves. The solution techniques can be expanded to the case where the excitation is a repetitive rectangular pulse string. However, there are other types of excitation for which the response of first-order circuits is not so simple. The most common of these is a sinusoid that is switched on at some specific time.

5.4.1.1 Sinusoidal AC Excitation The transient response of a first-order circuit to a sinusoid depends not only on the circuit component values, the amplitude of the source, and the time it is applied to the circuit, but also on the phase angle of the sinusoid at the time it is turned on. The range of possible responses is broad, and as a result, the technique used to solve the simpler problems given in Chapter 3 is not suitable. The simplest procedure is to use Laplace transforms, as in the following example.

EXAMPLE 5–7

What are $v_C(t)$ and $i(t)$ for the circuit shown in Figure 5–21(a) if $\theta = 0°$.

With a phase angle of 0°, the excitation is $e(t) = 10 \sin 3770t$. The first step is to transform the circuit. The *s*-plane representation of the resistance is the same as in the time domain. The capacitive reactance is transformed into $1/sC$, or $10^4/s$. The *s*-plane representation of the source is found using pair 4:

$$E(s) = 10\mathcal{L}[\sin \omega t] = \frac{10\omega}{s^2 + \omega^2} = \frac{37.70 \times 10^3}{s^2 + 14.21 \times 10^6}$$

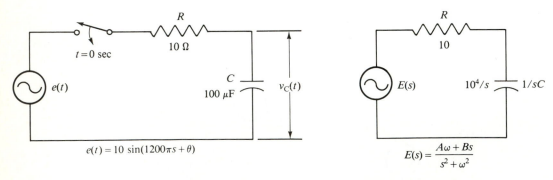

(a) Time domain representation   (b) *s*-plane representation

FIGURE 5–21   Circuit for Example 5–7

The s-plane circuit is shown in Figure 5–21(b). Using the voltage divider rule, we get the following expression for $V_C(s)$:

$$V_C(s) = I(s)Z_C(s) = \frac{E(s)Z_C(s)}{Z_t(s)}$$

So we have

$$V_C(s) = \frac{37.70 \times 10^3}{s^2 + 14.21 \times 10^6} \cdot \frac{10^4/s}{10 + 10^4/s} = \frac{37.70 \times 10^6}{(s^2 + 14.21 \times 10^6)(s + 1000)}$$

We can use the formula method Equation (4–39) to find the part of the solution due to the quadratic factor $s^2 + 14.21 \times 10^6$. Here, $a = 0$ and $\omega = 3770$. Thus,

$$Q(-a + j\omega) = \frac{37.70 \times 10^6}{s + 1000}\bigg|_{s = j3770} = \frac{37.70 \times 10^6}{j3770 + 1000}$$

$$= \frac{37.70 \times 10^6}{3900\angle 75.14°} = 9.666 \times 10^3 \angle -75.14°$$

Hence,

$$M = 9.666 \times 10^3$$

and

$$\theta = -75.14°$$

Using the formula, we obtain

$$f_1(t) = \frac{M}{\omega} \sin(\omega t + \theta) = \frac{9.666 \times 10^3}{3770} \sin(3770t - 75.14°)$$

$$= 2.564 \sin(3770t - 75.14°)$$

Partial fraction expansion provides the s-plane expression for the rest of the solution:

$$A = \frac{37.70 \times 10^6}{s^2 + 14.21 \times 10^6}\bigg|_{s = -1000} = 2.479$$

In the s-plane, this part of the solution becomes

$$F_2(s) = \frac{2.479}{s + 1000}$$

and, using transform pair 2, we get

$$f_2(t) = 2.479\epsilon^{-1000t}$$

The complete solution is

$$v_C(t) = 2.564 \sin(3770t - 75.14°) + 2.479\epsilon^{-1000t} \text{ V}$$

which is plotted in Figure 5–22. We can find the current directly in the time domain by means of the equation

$$i(t) = C\frac{dv}{dt}$$

relating current and voltage in a capacitor. Differentiating each term in the solution for the capacitor voltage, and multiplying by C, we get

$$i(t) = 0.9666 \cos(3770t - 75.14°) - 0.2479\epsilon^{-1000t} \text{ A}$$

In both results, the first term represents a constant-amplitude sinusoid with $\omega = 3770$ radians per second. This is the forced response, a function of the excitation and the circuit parameters. Since it continues at a constant amplitude, it is also called the *steady-state response*. The final term is the natural response, which is a function of the initial condition, the excitation, and the circuit. In the circuit presented here, it is also the transient response, since it decays to zero.

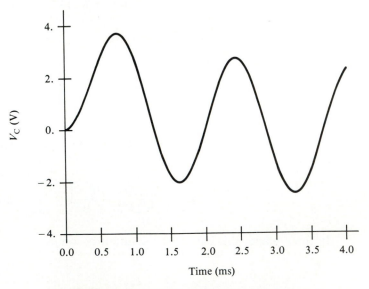

FIGURE 5–22   *Voltage response for Example 5–7*

The transient part of the response is the part that causes the total response to equal the initial condition at the output and then to approach the value of the forced component as time goes on. If the initial value is zero volts and the forced response is not zero, the natural response must be equal and opposite in sign to the forced response at $t = 0$ seconds. This can be seen in the preceding expressions for $v_C(t)$ and $i(t)$ and is a partial check on the results. Errors can cause both parts of the response to be incorrect, but the forced component can be checked by steady-state AC circuit techniques.

EXAMPLE 5–8   Find $v_C(t)$ for the circuit in Figure 5–21(a) if the phase angle θ is 45°.
Except for the excitation, the s-plane circuit is the same as before. Now the excitation is

$$e(t) = 10 \sin(3770t + 45°)$$

To transform this, we need to express the excitation as the sum of a sine wave and a cosine wave. Using Equation (5–2), we get

$$e(t) = 10 \sin 45° \cos 3770t + 10 \cos 45° \sin 3770t$$

which equals

$$e(t) = 7.071 \cos 3770t + 7.071 \sin 3770t$$

Using pairs 3 and 4, we get, for the s-plane source voltage,

$$E(s) = \frac{7.071(s + 3770)}{s^2 + 14.21 \times 10^6}$$

and then, using the voltage divider rule, we obtain

$$V_C(s) = I(s)Z_C(s) = \frac{E(s)Z_C(s)}{Z_t(s)}$$

or

$$V_C(s) = \frac{7.071(s + 3770)(1/100 \times 10^{-6}s)}{(s^2 + 14.21 \times 10^6)(10 + 1/100 \times 10^{-6}s)}$$

which simplifies to

$$V_C(s) = \frac{7071(s + 3770)}{(s^2 + 14.21 \times 10^6)(s + 1000)}$$

In this example,

$$Q(s) = \frac{7071(s + 3770)}{s + 1000}$$

and, as in the previous example, $\omega = 3770$ and $a = 0$. So

$$Q(-a + j\omega) = \frac{7071(s + 3770)}{s + 1000}\bigg|_{s = j3770} = \frac{7071(j3770 + 3770)}{1000 + j3770}$$

$$= 9.666 \times 10^3 \angle -30.14°$$

and

$$f_1(t) = 2.564 \sin(3770t - 30.14°)$$

To find the rest of the solution, we evaluate the constant

$$A = \frac{7071(s + 3770)}{s^2 + 14.21 \times 10^6}\bigg|_{s = -1000} = \frac{7071(2770)}{15.21 \times 10^6} = 1.288$$

and in the s-domain,

$$F_2(s) = \frac{1.288}{s + 1000}$$

Using transform pair 2, we get the time domain response:

$$f_2(t) = 1.288\epsilon^{-1000t}$$

The complete solution is

$$v_C(t) = 2.564 \sin(3770t - 30.14°) + 1.288\epsilon^{-1000t}$$

This is plotted in Figure 5–23. For the forced response, we have a sinusoid of the same frequency and amplitude as before, but shifted 45°. This is what we should expect, since the only change in the circuit from that of Example 5–7 was a 45° phase shift in the excitation. We can show this by performing a steady-state analysis on the circuit. The natural response, which is the part of the response that causes a transition from the initial state to the steady-state response, is somewhat changed in amplitude from the preceding example. However, its rate of decay is the same, since that is determined by the sizes of the circuit components, which are the same.

5.4.1.2 First-Order Circuits with Initial Conditions   When a circuit has energy stored in a capacitor or inductor at the start of a transient, the general character of the natural response is unchanged, except for its magnitude. The forced response

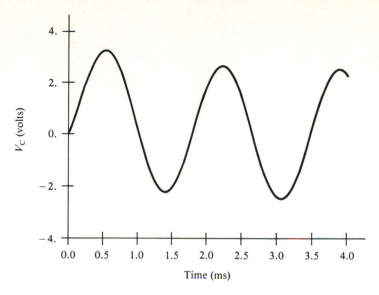

FIGURE 5–23 *Voltage response for Example 5–8*

will remain the same as it was for a circuit with zero initial conditions. With this in mind, we can continue the solution of the RL circuit with initial conditions started in Example 5–6.

EXAMPLE 5–9 In solving the RL circuit with initial conditions shown in Figure 5–17, we found that the s-plane expression for the current was

$$I(s) = \frac{5(s + 0.4)}{s(s + 0.2)}$$

Partial fraction expansion, as described in Chapter 4, can be used to find simpler expressions that can be inverted to get the current in the time domain using Tables 4–1 and 4–2. The expanded-fraction expression will be

$$I(s) = \frac{A}{s} + \frac{B}{s + 0.2}$$

Evaluating the numerators, we get

$$A = \left.\frac{5(s + 0.4)}{s + 0.2}\right|_{s = 0} = \frac{5(0.4)}{0.2} = 10$$

and

$$B = \left.\frac{5(s + 0.4)}{s}\right|_{s = -0.2} = \frac{5(-0.2 + 0.4)}{-0.2} = -5$$

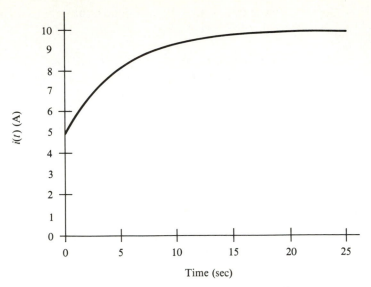

FIGURE 5–24 Plot of $i(t) = 10 - 5\epsilon^{-0.2t}$

Therefore,

$$I(s) = \frac{10}{s} - \frac{5}{s + 0.2}$$

Using transform pairs 1 and 2, we find the time domain response of the RL circuit with initial conditions to be

$$i(t) = 10 - 5\epsilon^{-0.2t}$$

The curve of this equation is plotted in Figure 5–24. The plot shows that the current starts at the initial condition value, 5 A, and rises exponentially to the steady-state or final value, 10 A.

5.4.2 Higher Order Circuits

When a system has two or more independent energy storage components, its responses are more complex than the previous examples. The degree of complexity depends on details of the circuit. A common circuit of this type has one resistance, one inductance, and one capacitance. The differential equation describing its response will accordingly have three terms, each involving a different order of derivative. The response equation in the s-plane will have three response terms also, each having a different power of s. The equation will have two roots, because the circuit has two independent energy storage elements. Note that the number of independent energy storage devices, the highest order derivative in the differential equation, the highest power of s in the characteristic equation, and the number of roots in the characteristic equation are equal.

5.4.2.1 Series RLC Circuits with DC Applied
A common, fairly simple second-order circuit is the series RLC circuit with DC suddenly applied at $t = 0$ seconds, as shown in Figure 5–25. We can transform this circuit to the s-plane circuit shown in Figure 5–26. If we want to find the current that flows as a result of the applied voltage, we can start by applying Kirchhoff's voltage law to the s-plane circuit. This gives

$$\frac{E_1}{s} = \left(sL + R + \frac{1}{sC}\right)I(s) \tag{5–18}$$

which can be solved for the current in the s-domain. The result is

$$I(s) = \frac{E_1/s}{sL + R + 1/sC}$$

We would prefer that the powers of s in the denominator be positive and the

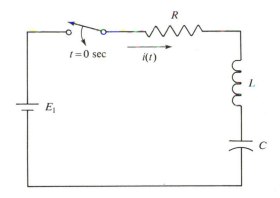

FIGURE 5–25 Series RLC circuit in the time domain

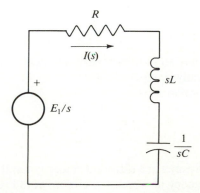

FIGURE 5–26 Series RLC circuit in the s-plane

coefficient of the highest power be one, for ease in factoring. Multiplying numerator and denominator by s/L, we get

$$I(s) = \frac{E_1/L}{s^2 + Rs/L + 1/LC} \tag{5-19}$$

We need to factor the denominator next. As described earlier, the roots may be real or complex and are equal to

$$s_1, s_2 = -\frac{b}{2} \pm \sqrt{\left(\frac{b}{2}\right)^2 - c}$$

or, in this context,

$$s_1, s_2 = -\frac{R}{2L} \pm \sqrt{\frac{R^2}{4L^2} - \frac{1}{LC}} \tag{5-20}$$

The characteristics of the roots thus depend on the values of the components and how they affect the expression under the radical. The roots are real and unequal if $R^2/4L^2 > 1/LC$. The solution in the time domain then consists of a sum of two exponentials with different time constants. If $R^2/4L^2 = 1/LC$, the expression under the radical is zero and the roots are real and equal. The solution is then a sum of two equal decaying exponentials, one of which is multiplied by t. The exponentials will have a decay rate that is intermediate between the rates for the two exponentials with real, unequal roots. As a result, the transient decays more rapidly. If $R^2/4L^2 < 1/LC$, the expression under the radical is negative. The roots are then complex conjugates, so the solution will be an exponentially decaying sinusoid. If there is no resistance in the circuit, the roots are imaginary and conjugate and the solution is a sinusoid.

The resistance required to give real, equal roots is called the *critical resistance* R_C. When $R = R_C$, the total response approaches the forced response more quickly than with other values of resistance. The ratio of the actual resistance to the critical resistance is called ζ, the damping ratio. That is,

$$\zeta = \frac{R}{R_C} \tag{5-21}$$

When $\zeta < 1$, the circuit has an exponentially damped sinusoidal response. When $\zeta = 1$, the circuit is critically damped.

The characteristic equation is

$$s^2 + \frac{Rs}{L} + \frac{1}{LC} = 0$$

We can identify the coefficients in a different, more general way. The constant $1/LC$ is the square of the frequency at which the circuit would resonate if there were no damping. We can call it ω_n, the *undamped resonant frequency*. In a second-order mechanical system, this would be equal to K/M, the spring constant divided by the mass (see Table 3–1). In the characteristic equation, the second term has a coefficient of R/L. When the circuit is critically damped, $R = R_C$ and the two terms under the radical in Equation (5–20) are equal. So

$$\frac{R_C^2}{4L^2} = \frac{1}{LC}$$

Solving for R_C, we obtain

$$R_C = 2\sqrt{L/C}$$

so that

$$R = \zeta R_C = 2\zeta\sqrt{L/C}$$

Substituting the right side for the coefficient of the second term of the characteristic equation, we get

$$s^2 + 2\zeta\omega_n s + \omega_n^2 = 0 \qquad (5\text{--}22)$$

This is the general form of the characteristic equation for a second-order system. As long as the damping ratio ζ and the undamped natural frequency ω_n are known, it can be used to find the general characteristics of a second-order system.

It can be seen that as the component values are varied, the responses change. It is worthwhile to see how this change occurs. If we let the resistance vary over a wide enough range, we can produce all of the responses just described.

EXAMPLE 5–10 In the circuit shown in Figure 5–25, what is the current if $L = 20$ henrys, $C = 5$ farads and $R = 16$ ohms? Let $e(t)$ be 10 V DC, switched on at 0 seconds.

The first step is to transform the circuit. The transformed circuit is shown in Figure 5–26. The current is

$$I(s) = \frac{0.5}{s^2 + 0.8s + 0.01} \qquad (5\text{--}23)$$

The denominator roots are

$$s_1, s_2 = -0.4 \pm \sqrt{0.16 - 0.01}$$

$$= -0.4 \pm 0.3873 = -0.0127, -0.7873$$

So

$$I(s) = \frac{0.5}{(s + 0.0127)(s + 0.7873)}$$

Using partial fraction expansion, we get

$$I(s) = \frac{A}{(s + 0.0127)} + \frac{B}{(s + 0.7873)}$$

where

$$A = \frac{0.5}{(s + 0.7873)}\bigg|_{s = -0.0127} = 0.6455$$

and

$$B = \frac{0.5}{(s + 0.0127)}\bigg|_{s = -0.7873} = -0.6455$$

Thus, after expansion, we have

$$I(s) = \frac{0.6455}{(s + 0.0127)} - \frac{0.6455}{(s + 0.7873)}$$

Using transform pair 2, we get, in the time domain,

$$i(t) = 0.6455\epsilon^{-0.0127t} - 0.6455\epsilon^{-0.7873t}$$

The current is thus the sum of a positive and a negative decaying exponential function of equal amplitude, but with widely different time constants. The curve is plotted in Figure 5–27. Note that the negative term, due to the second exponential

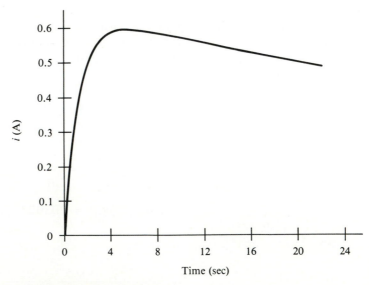

FIGURE 5–27   Plot of $i(t) = 0.6455\epsilon^{-0.0127t} - 0.6455\epsilon^{-0.7873t}$

function, decays much more rapidly than the positive term. It has some effect on the initial stages of the response, but after about 0.06 second, it has little effect. At that time, the response is essentially that due to the first term, a single decaying exponential. Because of this behavior, the exponential function with the slower time constant is sometimes called the *dominant term*.

EXAMPLE 5–11 Find the current if the resistor in the circuit shown in Figure 5–25 is reduced to 8 ohms.

With the resistance reduced to 8 ohms, R/L is now 0.4. This is the coefficient of the second term in the denominator of Equation (5–23). Thus, the current in the s-plane is now

$$I(s) = \frac{0.5}{s^2 + 0.4s + 0.01}$$

so the roots of the denominator term are

$$s_1, s_2 = -0.2 \pm \sqrt{(0.2)^2 - 0.01} = -0.2 \pm 0.1732$$
$$= -0.02680, -0.3732$$

and

$$I(s) = \frac{0.5}{(s + 0.02680)(s + 0.3732)}$$

Partial fraction expansion gives

$$I(s) = \frac{1.443}{s + 0.02680} - \frac{1.443}{s + 0.3732}$$

which transforms into

$$i(t) = 1.443\epsilon^{-0.0268t} - 1.443\epsilon^{-0.3732t}$$

The time constants of these exponential functions are more nearly equal than when the resistor was 16 ohms. The term with the shorter time constant has more effect on the shape of the sum than in Example 5–10. The current is plotted in Figure 5–28. Note that neither component of the response is dominant.

EXAMPLE 5–12 Find the current in the series RLC circuit when the resistor is 4 Ω.

This change in R changes R/L to 0.2. The current is then

$$I(s) = \frac{0.5}{s^2 + 0.2s + 0.01}$$

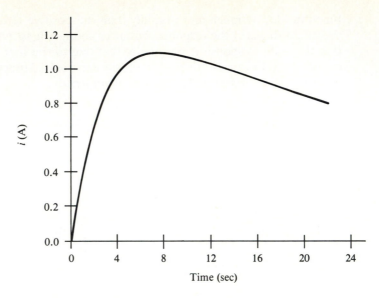

FIGURE 5–28   Plot of $i(t) = 1.443\epsilon^{-0.2268t} - 1.443\epsilon^{-0.5732t}$

and

$$s_1, s_2 = -0.1 \pm \sqrt{0.01 - 0.01} = -0.1, -0.1$$

There are thus two real, equal roots and

$$I(s) = \frac{0.5}{(s + 0.1)^2}$$

Using transform pair 6, we get, in the time domain,

$$i(t) = 0.5t\epsilon^{-0.1t}$$

This current is plotted in Figure 5–29. Its shape is generally similar to the curves in the previous two examples. The circuit is critically damped.

If it is necessary to find the voltage drop across the capacitor, the voltage divider rule can be applied. The capacitor voltage is

$$V_C(s) = \frac{Z_C}{Z_t} \cdot \frac{E}{s}$$

or

$$V_C(s) = \frac{1/sC}{Ls + R + 1/LC} \cdot \frac{E}{s} = \frac{E/LC}{s[s^2 + (R/L)s + 1/LC]}$$

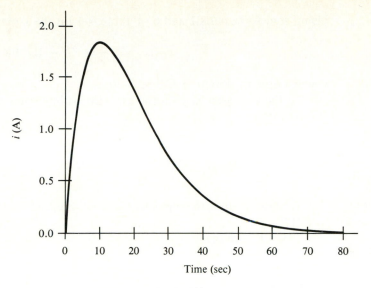

FIGURE 5–29 Plot of $i(t) = 0.5t\epsilon^{-0.1t}$

Entering the values for E, C, L, and R, we obtain

$$V_C(s) = \frac{0.1}{s(s^2 + 0.2s + 0.01)} = \frac{0.1}{s(s + 0.1)^2}$$

which is expanded to

$$V_C(s) = \frac{A}{s} + \frac{K_1}{(s + 0.1)^2} + \frac{K_2}{s + 0.1}$$

where

$$A = \frac{0.1}{(s + 0.1)^2}\bigg|_{s=0} = 0.1$$

$$K_1 = \frac{1}{0!} \cdot \frac{0.1}{s}\bigg|_{s=-0.1} = -1$$

and

$$K_2 = \frac{1}{1!} \cdot \frac{d}{ds}\left[\frac{0.1}{s}\right]\bigg|_{s=-0.1} = -10$$

The expanded fraction is thus

$$V_C(s) = \frac{10}{s} - \frac{1}{(s + 0.1)^2} - \frac{10}{s + 0.1}$$

Using transform pairs 1 and 6 of Table 4–1 on the s-plane equation, we get

$$v_C(t) = 10 - te^{-0.1t} - 10e^{-0.1t}$$

whose curve is plotted in Figure 5–30.

If the current is known, the capacitor voltage may be found more easily from the equation

$$v_C(t) = \frac{1}{C}\int_0^t i\,dt$$

EXAMPLE 5–13 What is the current in the series RLC circuit (Figure 5–25) if the resistance is changed to 2 Ω?

With this change, the coefficient R/L of s in the denominator is now 0.1, and

$$I(s) = \frac{0.5}{s^2 + 0.1s + 0.01}$$

We can apply transform pair 8 of Table 4–1 to this problem. If we let $\omega = 0.0866$ and $a = 0.05$, we obtain

$$I(s) = \frac{5.774(0.0866)}{(s + 0.05)^2 + (0.866)^2}$$

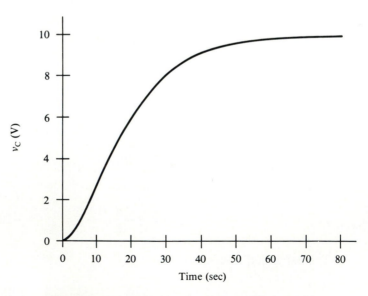

FIGURE 5–30 Plot of $V_C(t) = 10 - te^{-0.1t} - 10e^{-0.1t}$

In the s-plane, this becomes

$$i(t) = 0.5774\epsilon^{-0.05t} \sin 0.0866t$$

The curve of this equation is plotted in Figure 5–31. It is an exponentially damped sinusoid. The circuit is underdamped, its resistance being less than critical.

EXAMPLE 5–14 If the resistance of the circuit in the previous examples is zero ohms, the circuit will be undamped. What will the current then be?

We now have an LC series circuit. If $R = 0\ \Omega$, then $R/L = 0$ and the second coefficient will be zero. We then have

$$I(s) = \frac{0.5}{s^2 + 0.01}$$

Transform pair 4 of Table 4–1 can be used to obtain

$$i(t) = 5 \sin 0.1t$$

The application of a voltage step function to an LC circuit with zero resistance thus causes a constant-amplitude sinusoidal current to flow. The current is plotted in Figure 5–32.

5.4.2.2 Other Second-Order Circuits The series RLC circuit is not the only type of circuit whose operation is described by a second-order differential equation. A

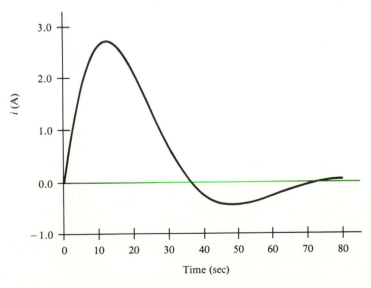

FIGURE 5–31 Plot of $i(t) = 0.5774\epsilon^{-0.05t} \sin 0.0866t$

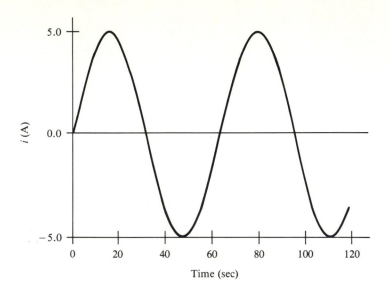

FIGURE 5–32 Plot of $i(t) = 5 \sin 0.1t$

parallel RLC circuit will be described by this type of equation also. Its response is similar to that of the series circuit if the voltages and currents are interchanged. It can have overdamped, critically damped, or damped oscillatory responses, depending on the value of the components.

The responses of mechanical systems that have both mass and springs are also described by higher order differential equations. Electric circuits and mechanical devices with multiple branches will have higher order responses if the energy storage devices cannot be combined. However, the range of responses that are possible may be more limited than with RLC circuits or mechanical BKM systems.

Second-Order Circuits with Nonoscillatory Responses Other circuits that have second-order responses are cascaded RL or RC filter circuits. One example is the RC circuit shown in Figure 5–33. In this circuit, the capacitors are neither in series nor in parallel, so they cannot be replaced by a single capacitor. The s-plane equivalent is shown in Figure 5–34. In this circuit, we can solve for V_2, the voltage across the capacitor, if we first find i_2. Then we can use Ohm's law to obtain

$$V_2(s) = \frac{I_2(s)}{sC_2}$$

The current divider rule can then be used to solve for I_2 in terms of the total current and the impedances. We get

$$I_2(s) = I_T(s) \cdot \frac{1/sC_1}{1/sC_1 + 1/sC_2 + R_2}$$

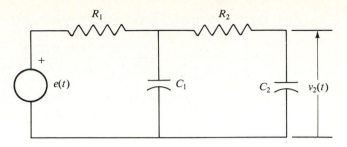

FIGURE 5–33 Second-order RC circuit in time domain

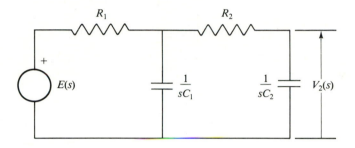

FIGURE 5–34 Second-order circuit in s-plane

and

$$E(s) = I_T(s)Z_T$$

The total impedance seen by the source is

$$Z_T = \left[\left(R_2 + \frac{1}{sC_2}\right)\middle\|\left(\frac{1}{sC_1}\right)\right] + R_1$$

Substituting this into the previous equation and solving for the total current from the source gives

$$I_T(s) = \frac{E(s)}{[(R_2 + 1/sC_2)\|1/sC_1] + R_1}$$

$V_2(s)$ is the voltage drop due to $I_2(s)$ flowing through capacitor C_2; so

$$V_2(s) = X_{C2}I_2(s) = \frac{1}{sC_2} \cdot \frac{1/sC_1}{1/sC_1 + 1/sC_2 + R_2} \cdot \frac{E(s)}{[(R_2 + 1/sC_2)\|1/sC_1] + R_1}$$

After some algebraic manipulation, we get

$$V_2(s) = \frac{1/R_1 R_2 C_1 C_2}{s^2 + [(R_1 C_2 + R_1 C_1 + R_2 C_2)/R_1 R_2 C_1 C_2]s + 1/R_1 R_2 C_1 C_2} \cdot E(s)$$

If the excitation is $U(t)$, a unit step function, $E(s)$ is $1/s$, so the voltage becomes

$$V_2(s) = \frac{1/R_1 R_2 C_1 C_2}{s^3 + [(R_1 C_2 + R_1 C_1 + R_2 C_2)/R_1 R_2 C_1 C_2]s^2 + (1/R_1 R_2 C_1 C_2)s}$$

One denominator factor is s. This gives a step function in the time domain. After factoring out s, we have a quadratic factor left. The two factors of this are found by

$$s_{2,3} = -\frac{R_1 C_2 + R_1 C_1 + R_2 C_2}{2R_1 R_2 C_1 C_2} \pm \sqrt{\left(\frac{R_1 C_2 + R_1 C_1 + R_2 C_2}{2R_1 R_2 C_1 C_2}\right)^2 - \frac{1}{R_1 R_2 C_1 C_2}}$$

The roots will be real and unequal if

$$\left(\frac{R_1 C_2 + R_1 C_1 + R_2}{2R_1 R_2 C_1 C_2}\right)^2 > \frac{1}{R_1 R_2 C_1 C_2}$$

or

$$(R_1 C_2 + R_1 C_1 + R_2 C_2)^2 > 4R_1 R_2 C_1 C_2$$

Squaring the expression in parentheses on the left, transposing the right-hand term to the left of the inequality, and collecting like terms we get

$$(R_1 C_1 - R_2 C_2)^2 + 2(R_1^2 C_1 C_2 + R_1 R_2 C_2^2)^2 + R_1^2 C_2^2 > 0$$

This expression is always positive, so the roots must be real and unequal, giving a response similar to that of an overdamped RLC circuit. Intuitively, the concept of energy being transferred back and forth between the energy storage components as in an RLC circuit seems less likely to be true in this circuit, since the phase shift between the currents into the two reactive devices will be less than 90°, rather than 180° as in a parallel RLC circuit.

Inductively Coupled Circuits   We can show how to find the transient response of a loosely coupled transformer by an example.

EXAMPLE 5–15   Find the voltage across the load resistor R_L in the circuit shown in Figure 5–35 if a unit step function voltage is applied at the input. The primary winding inductance is 5 H and its resistance is 7 Ω. The secondary has 10 H inductance and 10 Ω resistance. The mutual inductance is 4 H.

We can use the practical equivalent circuit shown in Figure 1–11. This is drawn in Figure 5–36. The transformed circuit is illustrated in Figure 5–37, where we have written the term for the voltage drop and indicated the polarity across each component. The preferred method of solution is mesh analysis.

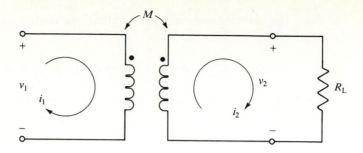

FIGURE 5–35   Circuit for Example 5–15

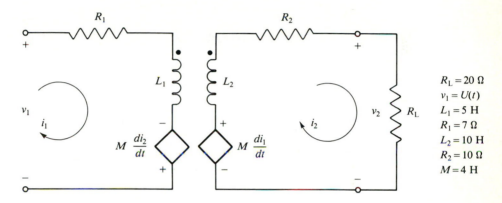

$R_L = 20\ \Omega$
$v_1 = U(t)$
$L_1 = 5\ \text{H}$
$R_1 = 7\ \Omega$
$L_2 = 10\ \text{H}$
$R_2 = 10\ \Omega$
$M = 4\ \text{H}$

FIGURE 5–36   Equivalent time domain circuit for circuit in Example 5–15

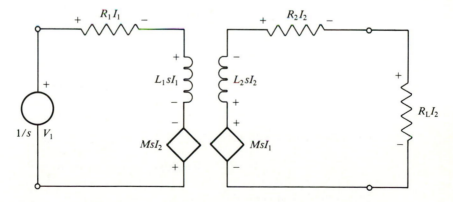

FIGURE 5–37   s-plane equivalent circuit for Example 5–15

The dot convention shows us that the mutual inductance terms should be negative. In the left-hand mesh, we find that

$$V_1(s) = (R_1 + L_1s)I_1(s) - MsI_2(s)$$

Summing up voltages around the other mesh, we get

$$0 = (R_2 + R_3 + L_2s)I_2(s) - MsI_1(s)$$

By Cramer's rule,

$$I_2(s) = \frac{MsV_1(s)}{(R_1 + L_1s)(R_2 + R_L + L_2s) - M^2s^2} \tag{5–24}$$

Substituting the given values into this and performing algebraic manipulations, we obtain

$$I_2(s) = -\frac{0.25}{s^2 + 13.75s + 13.13}$$

Factoring the quadratic term in the denominator, we have

$$s_1, s_2 = -\frac{13.75}{2} \pm \sqrt{\left(\frac{13.75}{2}\right)^2 - 13.13}$$

$$= -12.72, -1.032$$

So

$$I_2(s) = \frac{0.25}{(s + 12.72)(s + 1.032)}$$

Expanding this fraction, we get

$$I_2(s) = \frac{0.02139}{s + 1.032} - \frac{0.02139}{s + 12.72}$$

Transform pair 2 can be used to invert this, yielding

$$i_2(t) = 0.02139\epsilon^{-1.032t} - 0.02139\epsilon^{-12.72t}$$

Then Ohm's law gives

$$v_L(t) = i_1(t)R_L = 20i_2(t)$$

$$= 0.4279\epsilon^{-1.032t} - 0.4279\epsilon^{-12.72t}$$

The voltage across the secondary due to a unit step function voltage applied at the primary is thus the difference between two exponential functions. The curve will be somewhat similar in shape to that of the current shown in Figure 5–27. However, since the ratio of the two time constants is greater, the pole caused by the first term will be even more dominant.

5.4.3 Cases Where the Steady-State Response Is Zero

When the excitation becomes zero after a certain time, the response of a circuit will approach zero also, as long as there is some resistance in the circuit. In such cases, the component of the response caused by the excitation obviously cannot be called the steady-state response. Rather, it is called the *forced response,* which is an acceptable name for the steady-state response when one exists. The other part of the response is the natural response. Excitations that are decaying exponential functions or that have a decaying exponential multiplying factor are likely causes of such responses.

EXAMPLE 5–16   Figure 5–38 shows a series RL circuit with a decaying exponential source

$$e(t) = 5\epsilon^{-t}$$

Using the equations and procedures found in Sections 5.1 and 5.2 of this chapter, we can transform the circuit to that shown in Figure 5–39. Applying Ohm's law, we then get

$$E(s) = I(s)Z(s) = I(s)(R + sL)$$

and

$$\frac{5}{s+1} = I(s)(10 + 5s)$$

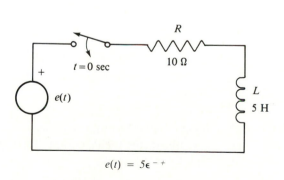

$$e(t) = 5\epsilon^{-t}$$

FIGURE 5–38 Series RL circuit with exponential source

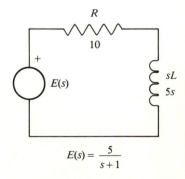

$$E(s) = \frac{5}{s+1}$$

FIGURE 5–39 *s*-plane equivalent of circuit of Figure 5–38

Solving for $I(s)$, we obtain

$$I(s) = \frac{5}{(s + 1)(5s + 10)} = \frac{1}{(s + 1)(s + 2)}$$

which can be expanded to

$$I(s) = \frac{K_1}{s + 1} + \frac{K_2}{s + 2}$$

where

$$K_1 = \left[\frac{1}{s + 2}\right]_{s = -1} = \frac{1}{-1 + 2} = 1$$

and

$$K_2 = \left[\frac{1}{s + 1}\right]_{s = -2} = \frac{1}{-2 + 1} = -1$$

So

$$I(s) = \frac{1}{s + 1} - \frac{1}{s + 2}$$

The time domain form may be found by using transform pair 2:

$$i(t) = \epsilon^{-t} - \epsilon^{-2t}$$

This function is plotted in Figure 5–40; the excitation approaches zero as time approaches infinity. The steady-state solution is zero.

The total solution has two parts. One is what we have previously called the natural response, since its time constant is $\frac{1}{2}$, the same as that of the circuit. The time constant of the second term in the response, ϵ^{-t}, indicates that this response is a result of the forcing function, so it is called the *forced response*. We can separate the two parts of the response in this case because they have different characteristics.

EXAMPLE 5–17   It is possible to have a forcing function that has the same decay rate as the natural response of the circuit. Figure 5–41 illustrates such a circuit. The s-plane equivalent is shown in Figure 5–42. Applying Ohm's law, we have

$$\frac{5}{s + 2} = I(s)(10 + 5s)$$

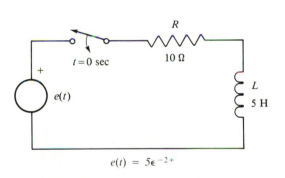

FIGURE 5-40 Plot of $i(t) = \epsilon^{-t} - \epsilon^{-2t}$

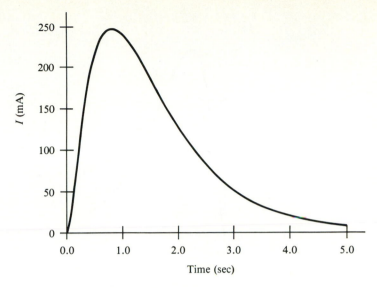

$$e(t) = 5\epsilon^{-2+}$$

FIGURE 5-41 Circuit for Example 5-17

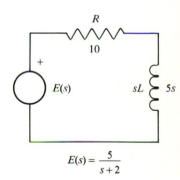

$$E(s) = \frac{5}{s+2}$$

FIGURE 5-42 *s*-plane equivalent of circuit in Figure 5-41

and

$$I(s) = \frac{1}{(s + 2)^2}$$

The denominator consists of two repeated factors, but we do not need to expand the fraction because transform pair 6 of Table 4–1 gives us the time domain function directly:

$$i(t) = t\epsilon^{-2t}$$

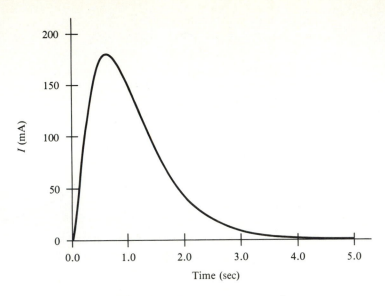

FIGURE 5–43 Plot of $i(t) = t\epsilon^{-2t}$

The curve of this equation is plotted in Figure 5–43. Because there is only one term on the right-hand side of the equals sign, we cannot identify or even separate the forced and the natural responses. Note that, as mentioned previously, $t\epsilon^{-nt}$ decays to zero as time approaches infinity.

PROBLEMS

5–1. Find the s-plane equivalent of the following ideal sources. (*Note*: If no step function multiplier is shown, all sources have $U(t)$ as a factor.)
   a. $e(t) = 5$
   b. $e(t) = 45 \sin(24\pi t + 85°)$
   c. $i(t) = 25\epsilon^{-45t}$
   d. $i(t) = 32 \cos(45\pi t - 79°)$

5–2. Find the equivalent sources for the following sources. All sources have zero output for $t < 0$ seconds.
   a. $i(t) = 92U(t - 5) \cos(28\pi t + 62°)$
   b. $i(t) = 25U(t - 10)$
   c. $e(t) = 29\delta(t - 2)$
   d. $e(t) = 45\epsilon^{-t} \sin(44t + 10°)$

5–3. Find the s-plane equivalent of the following components. Assume that no energy is stored.
   a. 35-Ω resistance
   b. 10-μF capacitance

    c. 30-mH inductance

    d. Loosely coupled transformer with $L_1 = 10$ H, $L_2 = 20$ H, $M = 5$ H, $R_1 = 5 \ \Omega$, $R_2 = 12 \ \Omega$

5–4. Find the s-plane equivalent of the following components. No energy is stored.

    a. 100-KΩ resistance

    b. 0.01-μF capacitance

    c. 125-H inductance

5–5. Find the series s-plane equivalent of the following components:

    a. 75-μF capacitance with $v_C(0^-) = 10$ V

    b. 100-mH inductance with $i_L(0^-) = 30$ mA

    c. 10-Ω resistance with $i_R(0^-) = 10$ A

    d. 25-pF capacitance with $i_C(0^-) = 30$ mA

5–6. Find the parallel s-plane equivalent of the following components:

    a. 82-Ω resistance with $v_R(0^-) = 10$ V

    b. 39-pF capacitance with $v_C(0^-) = 18$ V

    c. 10-H inductance with $i_L(0^-) = 100$ mA

    d. 7.5-μF capacitance in series with 29-H inductance, $v_C(0^-) = 0$ V and $i(0^-) = 125$ mA

5–7. Find both a Thévenin and a Norton equivalent for the charged capacitor shown in Figure 5P–1. Sketch the circuits.

5–8. Find both a Thévenin and a Norton equivalent for the energized inductor shown in Figure 5P–2. Sketch both circuits.

5–9. Transform the circuit shown in Figure 5P–3 if $v_C(0^-) = 0$ V. Sketch the transformed circuit.

5–10. Transform the circuit shown in Figure 5P–4 if $i_L(0^-) = 0$ A. Sketch the transformed circuit.

5–11. Transform the circuit shown in Figure 5P–5 if no components are energized at $t = 0$ seconds. Sketch the transformed circuit.

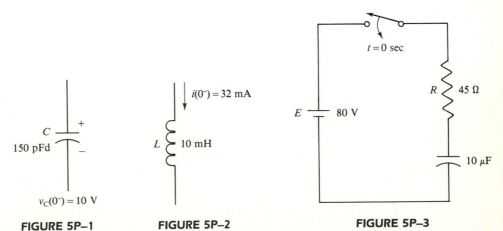

C
150 pFd

$v_C(0^-) = 10$ V

FIGURE 5P–1

$i(0^-) = 32$ mA

L 10 mH

FIGURE 5P–2

$t = 0$ sec

R 45 Ω

E 80 V

10 μF

FIGURE 5P–3

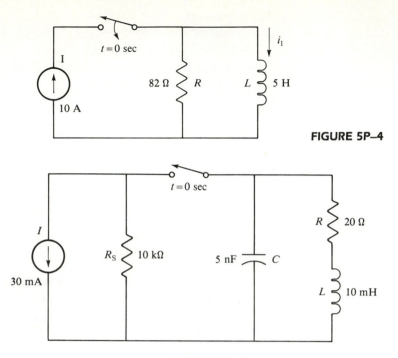

FIGURE 5P–4

FIGURE 5P–5

5–12. Transform the circuit shown in Figure 5P–6. No components are energized at $t = 0$ seconds. Sketch the transformed circuit.

5–13. Transform the circuit shown in Figure 5P–7 if $v_C(0^-) = 10$ V. Sketch the transformed circuit.

5–14. Transform the circuit in Figure 5P–8 if $v_C(0^-) = -45$ V. Sketch the transformed circuit.

5–15. In the circuit shown in Figure 5P–9, at $t = 0$ seconds the switch is opened. The capacitor is discharged at that time. Sketch the transformed circuit.

5–16. In the circuit shown in Figure 5P–10, the switch is closed at $t = 0$ seconds. Transform the circuit, and then sketch the transformed circuit.

5–17. Find the inverse transforms:

a. $F(s) = \dfrac{10}{s + 5}$

b. $I(s) = \dfrac{39s + 3.9 \times 10^3}{s^2 + 15.21 \times 10^6}$

c. $F(s) = \dfrac{12}{s^2 + 36}$

d. $I(s) = \dfrac{35s + 280}{s^2 + 16s + 89}$

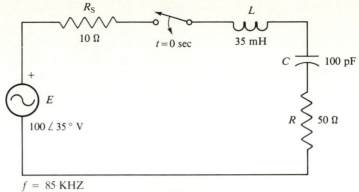

FIGURE 5P–6

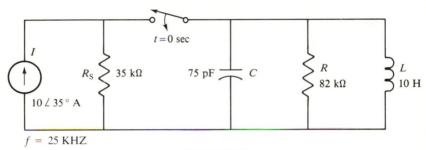

FIGURE 5P–7

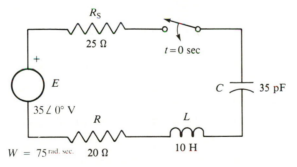

FIGURE 5P–8

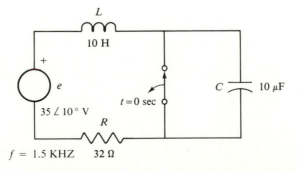

FIGURE 5P–9

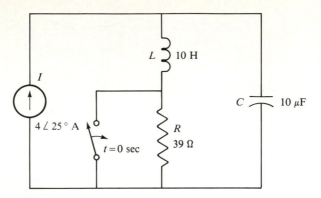

FIGURE 5P–10

5–18. Find $f(t)$ for the following:

a. $E(s) = \dfrac{25}{s^2 + 14s + 49}$

b. $F(s) = \dfrac{24s + 4046}{13s^2 + 3757}$

c. $E(s) = \dfrac{22}{s^3 + 64s^2 + 1024s}$

d. $F(s) = \dfrac{25s + 500}{s^2 + 16s + 3600}$

5–19. Solve for $i(t)$ in the circuit shown in Figure 5P–11, using Laplace transforms if $i_L(0^-) = 0$ A.

5–20. In Figure 5P–12, find $i_C(t)$, using Laplace transforms if $v_C(0^-) = 10$ V.

5–21. If the switch in Figure 5P–13 is closed for a long time before being opened at $t = 0$ seconds, find $v_C(t)$.

5–22. Find the expression for $v_L(t)$ in Figure 5P–14 if the switch is closed at $t = 0$ seconds.

5–23. In Figure 5P–15, if $v_C(0^-) = 13$ V, find $i_C(t)$.

5–24. In the circuit shown in Figure 5P–16, find $v_C(t)$ if $v_C(0^-) = 0$ V.

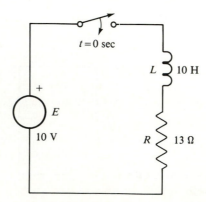

FIGURE 5P–11

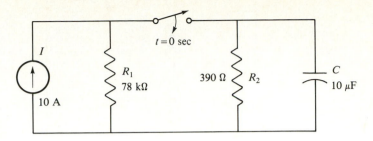

FIGURE 5P–12

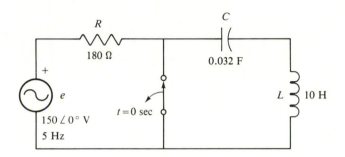

FIGURE 5P–13

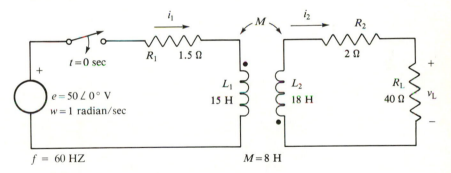

FIGURE 5P–14

5–25. Find $v_C(t)$, $i_L(t)$, $v_L(t)$ and $v_R(t)$ in Problem 5–23.

5–26. Find $i(t)$ and $v_R(t)$ in Problem 5–24. Find $v_L(t)$ using two methods: first find it directly using Laplace transforms, and then find it using the solution to Problem 5–24.

5–27. In Figure 5P–17, the switch has been closed for a long time when it is opened at $t = 0$ seconds. Find the current through the capacitor.

5–28. The circuit shown in Figure 5P–18 is energized by $e = 5U(t)\epsilon^{-2t}$ V. Find the voltage across the capacitor.

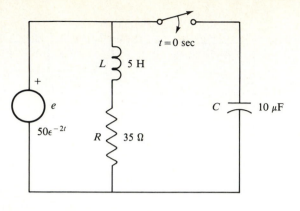

$t = 0$ sec

L　5 H

C　10 μF

e

$50\epsilon^{-2t}$

R　35 Ω

FIGURE 5P–15

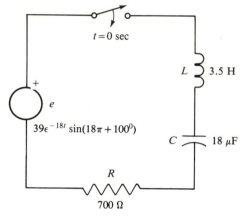

$t = 0$ sec

L　3.5 H

e

$39\epsilon^{-18t} \sin(18\pi + 100^0)$

C　18 μF

R

700 Ω

FIGURE 5P–16

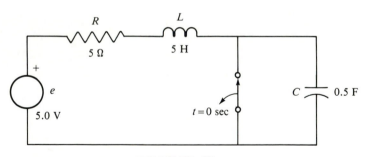

R

5 Ω

L

5 H

e

5.0 V

$t = 0$ sec

C　0.5 F

FIGURE 5P–17

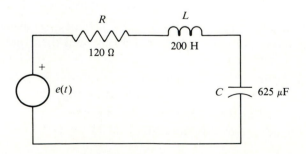

R

120 Ω

L

200 H

$e(t)$

C　625 μF

FIGURE 5P–18

5–29. The switch in Figure 5P–19 has been in position 1 for a long time before it is thrown to position 2 at $t = 0$ seconds. Solve for $i_2(t)$ for $t \geq 0$ seconds.

5–30. In the circuit shown in Figure 5P–20, the two switches have been in position 1 for a long time before they are thrown to position 2 at $t = 0$ seconds. Find $i(t)$ after $t = 0$ seconds.

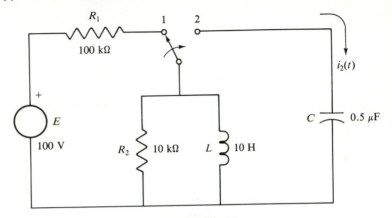

FIGURE 5P–19

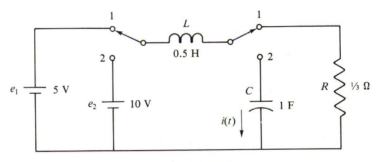

FIGURE 5P–20

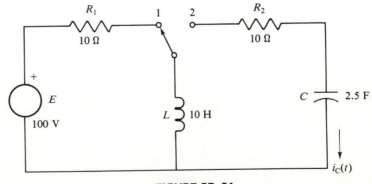

FIGURE 5P–21

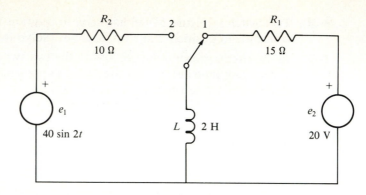

FIGURE 5P–22

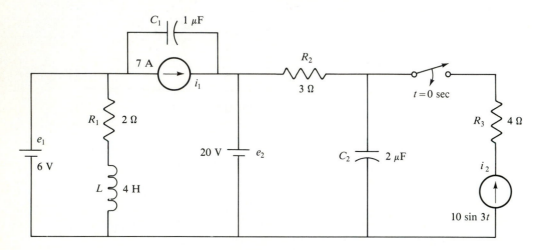

FIGURE 5P–23

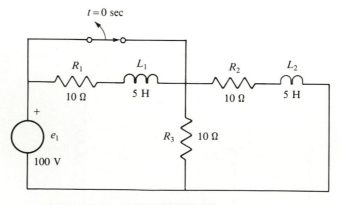

FIGURE 5P–24

5–31. In Figure 5P–21, the switch is thrown to position 2 at $t = 0$ seconds, after having been in position 1 for a long time. Find the capacitor current $i_C(t)$ for the time after the switch is thrown.

5–32. When $t < 0$ seconds, the circuit shown in Figure 5P–22 is in a steady-state condition with the switch in position 1. At $t = 0$ seconds, the switch is thrown to position 2. Solve for $i_1(t)$ for the time after $t = 0$ seconds. What are $v_L(t)$ and $v_R(t)$?

5–33. The switch in Figure 5P–23 has been open for a long time before $t = 0$ seconds, when it is closed. Transform the circuit. Sketch the transformed circuit.

5–34. In the circuit shown in Figure 5P–24, the switch is closed a long time before it is opened at $t = 0$ seconds. Draw the transformed circuit and write the mesh equations for it.

CHAPTER 6

Transfer Functions

6.1 OBJECTIVES

On completion of this chapter, you should be able to:

- Write transfer functions for electric circuits.
- Use transfer functions to find both the forced and the natural response of a circuit.
- Produce the pole-zero plot of a circuit and discuss how the placement of the poles and zeros affects the general character of the natural response.
- Evaluate the transfer function graphically over the s-plane.
- Evaluate the frequency response using the pole-zero plot.
- State the effects of pole location on stability.
- Find the impulse and step responses of a circuit.
- Find the responses of complex systems to sinusoids if the transfer functions of individual elements of the systems are known.
- Find the steady-state transfer function if the general transfer function is known.
- Draw the Bode plot for systems when their transfer functions are known.
- Sketch the straight-line approximations to the Bode plots.

6.2 INTRODUCTION

The transfer function has a number of uses in the analysis of electric circuits, both passive and active. It is a function of the individual elements in a circuit and bears a special relationship to the natural response. It can be useful in evaluating the response of circuits that have no energy stored in capacitors or inductors. It is equal to the response of a circuit to a unit impulse function. If it is evaluated along the $j\omega$ axis, it will give the AC response of a circuit. In this form, it is sometimes called the *steady-state transfer function*.

6.3 DEVELOPMENT OF THE TRANSFER FUNCTION

The transfer function of a circuit or system is defined as the ratio of the output to the input of the circuit or system when all three quantities are expressed in the s-plane. For the block in Figure 6–1, the formula giving the relationship between the input and the output is

$$Y(s) = T(s)X(s) \tag{6-1}$$

Here, $Y(s)$ is the response or output of the circuit and $X(s)$ is the input or excitation. $X(s)$ must be the only independent source in the circuit, including any initial conditions as well as other sources. $T(s)$ is the transfer function. We can express the transfer function explicitly by

$$T(s) = \frac{Y(s)}{X(s)} \tag{6-2}$$

Tables of transfer functions are available in the literature, particularly on automatic control systems.

For electric circuits, $X(s)$ and $Y(s)$ can be either voltage or current. Often, but not always, they are expressed in the same units. If so, the transfer function $T(s)$ is a voltage or current gain and has no units, or is nondimensional. However, if the excitation $X(s)$ is a current and the response $Y(s)$ a voltage, the transfer function is an impedance. It is then often called the *transfer impedance* or *transimpedance,* and its units are ohms. When the excitation is a voltage and the response is a current, the transfer function is an admittance, often called the *transfer admittance* or *trans-admittance*. Its units are then siemens.

The excitation and response used to define the transfer function are generalized exponential functions e^{st} or are closely related to these functions. Such functions include both real and complex exponential functions.

The step function and the impulse function are limiting cases of real exponential functions. The unit step function equals ϵ^0 for $t > 0$ seconds. The unit impulse function could be expressed as the limit of $\epsilon^{-\sigma t}$ as σ approaches infinity, or as the function approaches ϵ^∞, as described in Chapter 2.

We have seen in Chapter 2 that the imaginary exponents represent sinusoids. The unit ramp function described there is the integral of the unit step function, so it is a member of this group of exponentials.

FIGURE 6–1 Transfer function block

The definition of the transfer function requires that the network components be linear, time invariant, and lumped. Linearity implies that the amplitude of any response is directly proportional to the input. If all components are linear, then

$$|Y(s)| = k|X(s)| \qquad (6\text{--}3)$$

must be true for the system.

With linear components, superposition will apply. That is, if an input $X_1(s)$ produces an output equal to $Y_1(s)$ and $X_2(s)$ produces $Y_2(s)$, then when the sum $X_1(s) + X_2(s)$ is applied to the input, the output is $Y_1(s) + Y_2(s)$.

Real components are rarely linear for all ranges of inputs. For instance, if the amplitude is high enough, a resistor will heat up, changing its resistance. In most cases, electrical components are linear in their normal operating ranges. An exception is ferromagnetic devices. Even with these, however, a reasonable approximation to a linear model is possible.

The characteristics or parameters of a time-invariant component do not change with time. Most components in electric circuits meet this condition. The effects of those that do vary with time, such as tuning components in scanning radio receivers, can often be evaluated at several points across their range, as if they were fixed at those values. The transient characteristics can be evaluated for each value, to furnish both extreme and intermediate cases.

Lumped components are considered to have their impedance characteristics lumped at one location in the circuit. As described in Chapter 1, individual resistors, capacitors, and inductors can be considered to consist of several separate lumped components.

In transmission lines, resistance, capacitance, and inductance are distributed along the line. Thus, ordinary differential equations will not describe the responses exactly. Instead, partial differential equations, involving differentiation with respect to both time and distance along the line, are used for complete results. Also, lumped constant approximations can be used to give reasonably accurate equivalent circuits for transmission lines.

If the excitation and response in Equation (6–2) are of the generalized exponential group, then, from Tables 4–1 and 4–2, their transforms will be polynomials in s; that is,

$$T(s) = \frac{N(s)}{D(s)} \qquad (6\text{--}4)$$

Expressing the polynomials in more detail, we have

$$T(s) = \frac{a_0 s^n + a_1 s^{n-1} + \cdots + a_{n-1}s + a_n}{b_0 s^m + b_1 s^{m-1} + \cdots + b_{m-1}s + b_m} \qquad (6\text{--}5)$$

For real systems, n is equal to or less than m.

Any of the circuit analysis techniques given in earlier chapters may be used to find the transfer function of a circuit. The first step is to obtain the Laplace transform of the circuit components and variables. Then, we apply the proper circuit analysis method to find the output as a function of the input.

EXAMPLE 6–1   If $v_2(t)$ in the series RC circuit in Figure 6–2 is the output when the input is $v_1(t)$, what is the transfer function of the circuit?

The first step in obtaining the transfer function is to transform the circuit to the form shown in Figure 6–3. Then the voltage divider rule provides a simple method of solution:

$$V_2(s) = \frac{1/sC}{R + 1/sC} V_1(s)$$

Next, since the desired transfer function is the output $V_2(s)$ divided by the input $V_1(s)$, we obtain

$$T(s) = \frac{V_2(s)}{V_1(s)} = \frac{1/sC}{R + 1/sC}$$

This transfer function is a ratio of voltages, so it is dimensionless.

Later, we will see that it is often better to express the transfer function using only positive powers of s and with unity coefficients for the terms with the highest powers of s in both numerator and denominator. In this manner, we would have:

$$T(s) = \frac{1/RC}{s + 1/RC}$$

EXAMPLE 6–2   If, in the RLC circuit shown in Figure 6–4, the input is $v_1(t)$ and the output is $i_2(t)$, what is the transfer function $T(s) = I_2(s)/V_1(s)$?

The transformed circuit is shown in Figure 6–5. We can find the current by first applying the voltage divider rule to determine $V_C(s)$. We have

$$V_C(s) = \frac{(Z_C\|Z_R)}{Z_L + (Z_C\|Z_R)} V_1(s) \text{ V}$$

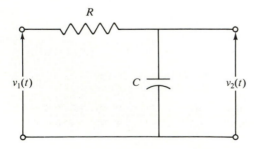

FIGURE 6–2   Time domain RC circuit

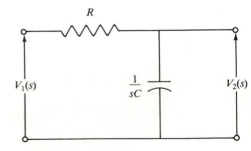

FIGURE 6–3   *s*-plane RC circuit equivalent to circuit of Figure 6–2

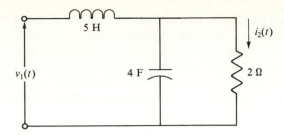

FIGURE 6–4 Time domain RLC circuit

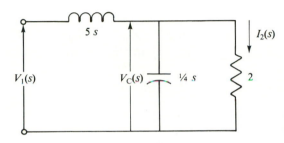

FIGURE 6–5 s-plane RLC circuit equivalent to circuit of Figure 6–4

where

$$Z_C \| Z_R = \frac{(1/4s)2}{(1/4s) + 2} = \frac{1}{8s + 2} \, \Omega$$

So

$$V_C(s) = \frac{1/(8s + 2)}{5s + [1/(8s + 2)]} V_1(s) = \frac{1}{40s^2 + 10s + 1} \text{ V}$$

Ohm's law then gives

$$I_2(s) = V_C(s)R = \frac{2}{40s^2 + 10s + 1} V_1(s) \text{ A}$$

Since the transfer function is a current divided by a voltage, it is an admittance and its units are siemens. The transfer admittance is thus

$$T(s) = \frac{I_2(s)}{V_1(s)} = \frac{2}{40s^2 + 10s + 1} = \frac{0.05}{s^2 + 0.25s + 0.025} \text{ S}$$

6.3.1 Poles and Zeros

The polynomials in the numerator and denominator of the transfer function given in Equation (6–5) can be factored into first-order terms in the form

$$T(s) = \frac{K(s - z_1)(s - z_2)\cdots(s - z_n)}{(s - p_1)(s - p_2)\cdots(s - z_m)} \tag{6–6}$$

K is an amplitude or gain factor that is equal to a_0/b_0 in Equation (6–5). When $s = z_i$, the numerator term $s - z_i$ is equal to zero. Thus, at the point z_i in the s-plane, the transfer function will equal zero. This point is said to be a *zero* of the transfer function. Similarly, when $s = p_i$, the corresponding denominator term is zero and the transfer function will be infinite at p_i in the s-plane. This point is called a *pole* of the transfer function. Poles and zeros are sometimes called *singularities* or *singular points*.

There can be poles and zeros at any point in the s-plane. That is, they could be on the axes, either real or imaginary, or they could be complex and off the axes. As indicated earlier, if the poles and zeros are imaginary or complex, they will occur in complex conjugate pairs. The location of the poles and zeros of a transfer function gives us information about the dynamic characteristics of a circuit.

EXAMPLE 6–3 Find the poles and zeros of the transfer function

$$T(s) = \frac{s + 5}{s^2 + 4s + 8}$$

The quadratic factor in the denominator can be factored, giving

$$T(s) = \frac{s + 5}{(s + 2 + j2)(s + 2 - j2)}$$

The constants in each of the factors give the locations of the singularities. The numerator factor, $s + 5$, indicates a zero on the real axis at $s = -5$. The complex conjugate factors in the denominator show a complex conjugate pole pair at $s = -2 - j2$ and $s = -2 + j2$.

6.3.2 Output Functions and Natural and Forced Responses

Another system function is the response function

$$Y(s) = T(s)X(s)$$

given in Equation (6–1). The equation says that the response equals the product of the transfer function and the excitation. As mentioned earlier, inspection of Table 4–1 shows that the s-plane expressions for exponential excitation have polynomial factors in s in their denominators. This means that excitation factors produce poles in response functions.

From the preceding examples, we see that the transfer function can have polynomials in both its numerator and its denominator. Accordingly, it can have both poles and zeros. The product of the excitation and the transfer function is called the *response function*. Since it can consist of one polynomial divided by another polynomial, it can also have poles and zeros.

The response function should have the poles and zeros of the transfer function plus the poles due to the excitation. It may be that one or more of the poles of either the transfer function or the excitation function coincide with a zero of the other function. If so, they will cancel each other out and will not appear in the output. Such cancellation is to be expected, considering that a zero in either function means that either a specific excitation does not occur or the system does not pass that frequency. In either case, the frequency in question will not be in the output.

The output function will include the Laplace transforms of the forced and natural responses. In Section 3.4, we pointed out that the forced response includes frequencies present in the excitation only. As a result, the poles of the excitation function will produce the forced response. If the excitation has a pole, that pole will appear in the output function, unless it is canceled by a zero at the same point in the s-plane. The natural response will be determined primarily by the transfer function, which, in turn, is determined by the circuit components. The natural response will be part of any response function, regardless of what the excitation is. Its magnitude will depend on the amplitude of the excitation.

6.3.3 Pole-Zero Plots

A plot of the poles and zeros for the transfer function of a typical circuit is shown in Figure 6–6. The horizontal dimension is σ, the real part of s. The imaginary part

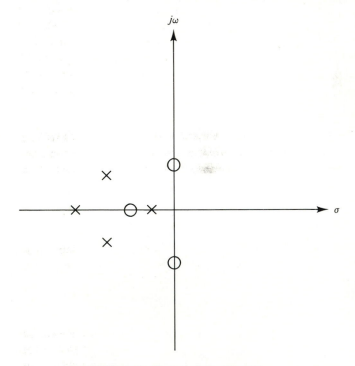

FIGURE 6–6 Typical pole-zero plot

of s, $j\omega$, is plotted vertically. In the figure, pole locations are marked by the letter X and zeros by the numeral 0. This is standard notation for pole-zero plots. We will see later that such a plot can be used to show everything of significance about the input-output relationship of the circuit, except gain factors and phase shifts. Poles and zeros that are off the real axis are found only as conjugate pairs.

EXAMPLE 6–4   Plot the poles and zeros of

$$T(s) = \frac{s^2 + 5s + 4}{s^3 + 2s^2 + 2s}$$

We can factor s out of the denominator, leaving quadratic factors in both the numerator and the denominator. Factoring these, we then find that the transfer function is

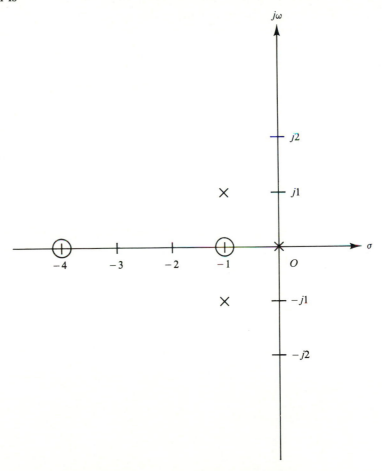

FIGURE 6–7   Poles and zeros for Example 6–4

$$T(s) = \frac{(s + 4)(s + 1)}{s(s + 1 + j1)(s + 1 - j1)}$$

There are thus two zeros on the negative real axis, at $s = -1$ and $s = -4$. Also, there is a pole at the origin and a conjugate pair of poles at $s = 1 \pm j1$. The three poles and two zeros of this transfer function are plotted in Figure 6–7.

6.3.4 Significance of Locations of Poles and Zeros

The poles of a circuit function indicate the locations of peak values of the function, and the zeros show where the function is zero. If the circuit function is a transfer function, the poles show which input functions produce strong responses and the zeros show which functions produce zero output. We can identify the time domain responses that correspond to the s-plane locations of the poles through the use of Tables 4–1 and 4–2. A plot of poles for typical systems is shown in Figure 6–8.

In Table 4–1, transform pair 2 indicates that a pole at $s = -a$ on the real axis represents a decaying exponential function ϵ^{-at} in the time domain. As the pole

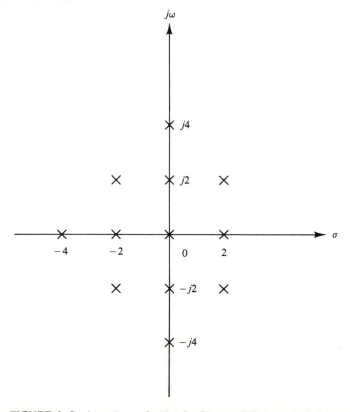

FIGURE 6–8   Locations of poles for Figures 6–9 through 6–16

moves away from the origin, the magnitude of the exponent increases. This means that as poles move away from the origin, the decay rate increases. This relationship is illustrated in Figures 6–9 and 6–10.

When the pole approaches the origin, the exponent and the decay rate approach zero. The time domain function is then a step function, as shown in Figure

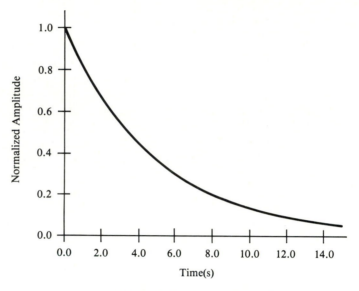

FIGURE 6–9   Pole at $s = -0.2$

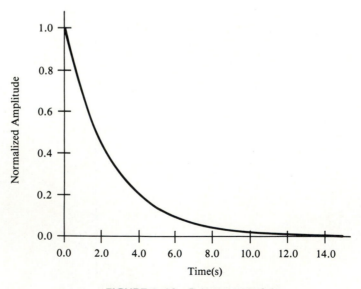

FIGURE 6–10   Pole at $s = -0.4$

6–11. If the pole is to the right of the $j\omega$-axis, the exponent in the time function is positive. This produces a growing exponential ϵ^{at}, which is illustrated in Figure 6–12. There is a continuous change from a rapid exponential decay for large negative exponents, through zero decay when $a = 0$, to a rapid exponential growth for large positive values of a.

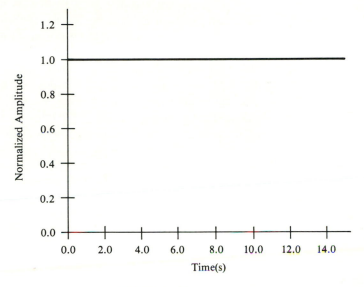

FIGURE 6–11   Pole at $s = 0$

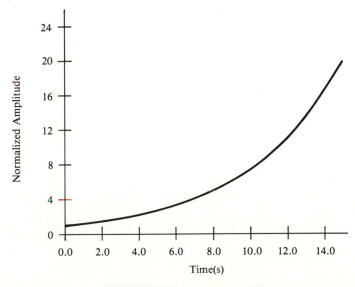

FIGURE 6–12   Pole at $s = 0.2$

Singularities having imaginary components occur only in conjugate pairs. That is, if a function has $s + j\omega$ as a factor in its denominator, $s - j\omega$ will also be a factor in the denominator. Multiplying these two factors together yields $s^2 + \omega^2$. Transform pairs 3 and 4 of Table 4–1 show that this is a denominator factor for sinusoidal signals. As the magnitude of the radian frequency, ω, increases, poles move away from the σ- or horizontal axis along the $j\omega$-axis. Figure 6–13 shows the sinusoidal response associated with a pole pair comparatively far from the origin on the $j\omega$-axis. If the poles are closer to the origin, the response is a lower frequency sinusoid, as shown in Figure 6–14. For a pole at the origin, the response is the constant-amplitude step function shown in Figure 6–11.

Poles that are on neither axis will also occur only in complex conjugate pairs. The denominator factors giving rise to these poles will be of the form $s - a - j\omega$ and $s - a + j\omega$. These poles might be expected to have characteristics of poles on both axes, because they have both real and imaginary offsets.

If we multiply the preceding complex pair, we get $(s - a)^2 + \omega^2$. Transform pairs 7 and 8 of Table 4–1 have such a denominator factor, indicating that complex conjugate poles represent exponentially decaying or growing sinusoids. The rate of decay or growth is proportional to the distance from the $j\omega$-axis, and the frequency is proportional to the distance from the σ-axis. This relationship is shown in Figures 6–15, 6–14, and 6–16. As the poles move from a location to the left of the imaginary axis to a location at the right, the envelopes of the responses change from a decaying exponential through a constant amplitude to a growing exponential.

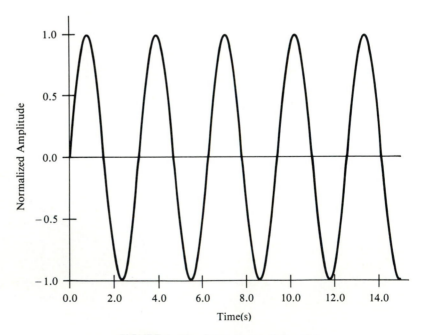

FIGURE 6–13 Poles at $s = j0.4, -j0.4$

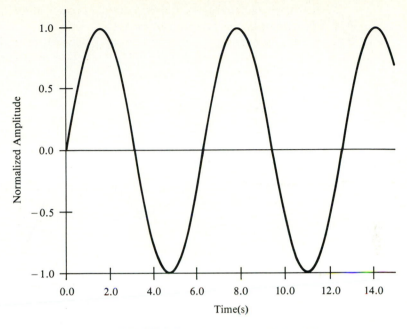

FIGURE 6–14   Poles at $s = j0.2, -j0.2$

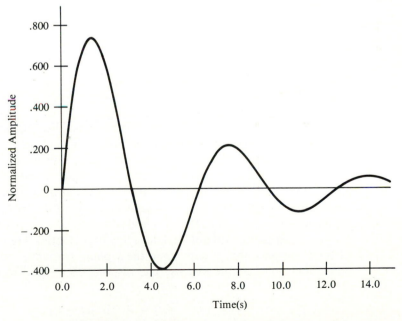

FIGURE 6–15   Poles at $s = -0.2 + j0.2, -0.2 - j0.2$

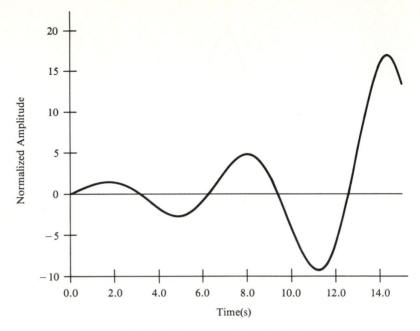

FIGURE 6–16   Poles at $s = 0.2 + j0.2, 0.2 - j0.2$

The locations of the poles of a transfer function in the s-plane represent the character of the natural response of the circuit. The locations of the poles, however, give neither the actual amplitude of the response nor the relative amplitudes of different factors in the response. Inspection of Figures 6–9 through 6–16 shows that as the distance of a pole from either axis increases, the time function due to the pole has steeper slopes. If the distance parallel to the real axis increases, the decay rate or growth rate increases. If the pole is moved vertically, the frequency of the sinusoid that is represented increases.

For poles to the right of the vertical axis, the function becomes a growing exponential. This function increases without limit; therefore, the system is unstable. For a pole on the vertical axis, the system is intermediate between stable and unstable. This condition is called *marginal stability*. If such a system is excited by a signal at the same point on the $j\omega$-axis, the output function will have two coincident pole pairs. This situation is somewhat similar to one described in Chapter 4 for real, repeated roots. Like that, the response will have a factor with a constant multiplier, but here that factor will not be a decaying exponential. Figure 6–17 shows the type of time response that occurs when the s-plane response function has two pairs of coincident poles on the $j\omega$-axis. The response is of the form

$$f(t) = At \, \sin(\omega t + \theta)$$

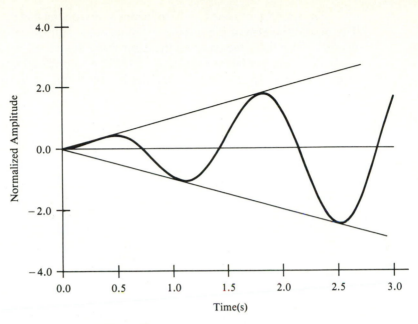

FIGURE 6–17   Two coincident pole pairs on $j\omega$-axis

In Chapter 5, the following solution for the roots of the characteristic equation of a series RLC network was given:

$$s_1, s_2 = -\frac{R}{2L} \pm \sqrt{\left(\frac{R}{2L}\right)^2 - \frac{1}{LC}} \tag{5–21}$$

This expression has a negative real part, indicating that the poles are on the left half of the s-plane. The only way it can represent a pole on the right half-plane is for $-R/2L$ to be positive. Self-inductance cannot be negative, but with the aid of amplifying devices, we can simulate negative resistance. This means that passive circuits cannot be unstable, while active circuits can.

To have a marginally stable system, the resistance must be zero. But inductors and capacitors always have some resistance, so passive circuits cannot be marginally stable. With active devices, the positive resistance of the circuit components can be compensated for by a negative resistance produced by the amplifier, so that the total resistance is zero.

**EXAMPLE
6–5**
Find the locations of the poles and zeros for the transfer function

$$T(s) = \frac{(s + 5)(s^2 + 4s + 8)}{(s + 7)(s^2 + 25)(s^2 + 10s + 36)}$$

Discuss how the singular points affect the character of the response of the system.

The numerator factor $s + 5$ indicates a zero on the negative real axis at $s = -5$. The quadratic term in the numerator indicates a pair of complex conjugate zeros. Factoring this term, we find that these zeros are at $s = -2 \pm j2$. There is a pole due to the denominator factor $s + 7$ at $s = -7$. The second denominator factor, $s^2 + 25$, indicates a pair of conjugate poles. Setting the term equal to zero, we have

$$s^2 + 25 = 0$$

and

$$s = \pm j5$$

These poles are on the $j\omega$-axis.

Factoring the final quadratic factor, we obtain

$$s_1, s_2 = -\frac{10}{2} \pm \sqrt{\left(\frac{10}{2}\right)^2 - 36} = 5 \pm j3$$

This gives us a pair of complex conjugate poles at those points.

The factored transfer function is

$$T(s) = \frac{(s + 5)(s + 2 + j2)(s + 2 - j2)}{(s + 7)(s + j5)(s - j5)(s + 5 + j3)(s + 5 - j3)}$$

The zeros do not coincide with any of the poles, so they cannot cancel any. If they are close to a pole, they will reduce the amplitude of any response caused by excitations close to that pole. This is so because the magnitude of the zero term will be small for an excitation at any point nearby. Approaching the pole itself, the magnitude of the pole term will be very large, of course. As mentioned earlier, the poles will have much more of an effect on the response; specifically, they will produce the natural response of the system. For this transfer function, we expect to find output terms like $A\epsilon^{-7t}$, $B \sin(5t + \theta)$ and $A\epsilon^{-5t} \sin(3t + \phi)$ in the natural response. The magnitudes and the phase angles are unknown.

6.3.5 Evaluation of Transfer Functions

If we have the transfer function in factored form, that is,

$$T(s) = \frac{K(s - z_1)(s - z_2) \cdots (s - z_n)}{(s - p_1)(s - p_2) \cdots (s - z_m)}$$

we can readily evaluate it graphically. This may prove to be of value in estimating the effect of changes in the design of a circuit to improve its operation. With scientific calculators readily available, the transfer function is easier to evaluate by computation. Measurements made on a reasonably careful sketch of the vectors could furnish a check on computations done with a calculator.

If there is a pole at location p_1 and excitation at point s_1, the vector $\mathbf{s}_1 - \mathbf{p}_1$ is equal to the vector $\mathbf{s}_1$ minus the vector $\mathbf{p}_1$, as shown in Figure 6–18. The preceding transfer function may then be evaluated for the system response to any exponential excitation by substituting the s-plane representation of the excitation for s in each factor in the numerator and in each in the denominator. Graphically, each factor is a vector going from the pole or zero to the location of the excitation in the s-plane.

EXAMPLE 6–6

Evaluate the following transfer function for a signal at $s = 0$, that is, a DC signal:

$$T(s) = \frac{s^2 - 2}{(s^2 + 6s + 25)(s + 4)}$$

The numerator and denominator factors should be factored first. The factored transfer function is

$$T(s) = \frac{(s + j2)(s - j2)}{(s + 3 + j4)(s + 3 - j4)(s + 4)}$$

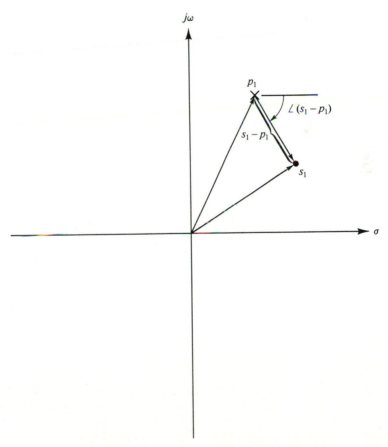

FIGURE 6–18 Graphical evaluation of vectors

The pole-zero plot is given in Figure 6–19. The excitation at the origin is marked by a solid dot. The vectors from the zeros to the excitation at $\pm j2$ are $2\angle 90°$ and $2\angle 270°$. The vectors from the poles at $-3 \pm j4$ are $5\angle 126.9°$ and $5\angle 233.1°$. The vector from the pole at -4 is $4\angle 180°$. Substituting these values into the transfer function, we get

$$T(0) = \frac{(2\angle 90°)(2\angle 270°)}{(5\angle 126.9°)(5\angle 233.1°)(4\angle 180°)} = -0.040\angle 0°$$

The magnitude of the transfer function is the product of the vectors from the excitation to the zeros of the transfer function divided by the product of the vectors from the excitation to the poles. The angle of the transfer function is the sum of the angles of the zeros minus the sum of the angles of the poles. In this case, the excitation is DC, so the result should be a scalar. We can express the results as $T(0) = -0.40$.

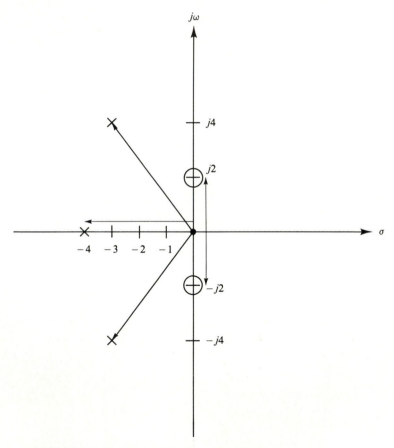

FIGURE 6–19 Graphical evaluation of response from pole-zero plot

With the foregoing technique, it is possible to find the frequency response of a circuit if the transfer function is known. Graphically, the excitation point is placed at a number of points along the imaginary axis, and at each point the magnitude and phase of the vectors to the poles and zeros are measured. These are then used to evaluate the transfer function as a function of frequency. In evaluating the response with a calculator, the excitation locations are substituted for s in the transfer function.

EXAMPLE 6–7   Find the response of the circuit having the transfer function

$$T(s) = \frac{s + 5}{(s + 3)(s + 8)}$$

for sine waves of 1-volt amplitude at the frequencies $\omega = 2$ and $\omega = 6$. The results are to be used in producing plots of amplitude and phase versus frequency for the circuit.

For $\omega = 2$, corresponding to a frequency of 2 radians per second, or 0.3183 hertz, we substitute $j2$ for s in the transfer function:

$$T(j2) = \frac{j2 + 5}{(j2 + 3)(j2 + 8)} = \frac{5.385\angle 21.80°}{(3.606\angle 33.69°)(8.246\angle 14.04°)}$$

$$= 0.1811\angle -25.93°$$

When $\omega = 6$ radians per second, or $f = 0.9549$ hertz, we substitute $j6$ for s in the transfer function, obtaining

$$T(j6) = \frac{j6 + 5}{(j6 + 3)(j6 + 8)} = \frac{7.180\angle 50.19°}{(6.708\angle 63.43°)(10.00\angle 36.87°)}$$

$$= 0.1070\angle -0.11°$$

To find the response to a specific voltage, we have to use the equation defining the transfer function, that is,

$$Y(s) = T(s)X(s)$$

The magnitude of each input is given as 1 V. The phase shift was not given, but since we are to use the data in frequency response plots, we can assume that the input phase angle is 0°. Accordingly, the values for $X(s)$ are 1 V$\angle 0°$ for each input. So

$$Y(j2) = T(j2)X(j2) = (0.1813\angle -25.93°)(1\angle 0°)$$

$$= 0.1811\angle -25.93° \text{ V}$$

and

$$Y(j6) = T(j6)X(j6) = 0.1070\angle -0.11° \text{ V}$$

These are the calculated values of two points on the frequency response curve for the system. By calculating the values of a number of points across the entire band, we would get the complete frequency response.

6.4 RESPONSES TO SPECIFIC INPUT FUNCTIONS

Often, special waveforms are used to evaluate electrical devices. For example, as shown in the preceding section, the sine wave can be used to find the frequency response of amplifiers. Many times, it is necessary to find the amplitude of the response across a specific frequency band. In many cases, particularly when the circuit is part of a control system, it is necessary to measure the phase shift as a function of frequency as well.

Sine wave frequency response testing is often the best method for evaluating the operation of electronic circuits. It is a comparatively slow method, however. A quicker method of evaluating the general characteristics of a system is to find the response to a step function or an impulse function. These functions have broad spectra, so a single step or impulse will provide a thorough test of a system.

Step and impulse functions also have discontinuities, which makes it impossible to generate them exactly. However, they may be approximated with enough accuracy to test specific devices if the discontinuities are replaced by sufficiently fast rises and falls. To simulate an impulse function, both the rise and fall must occur very quickly. In testing control systems or pulse circuits, testing with steps or impulses produces results that are more directly usable without extensive computation or interpretation than frequency response testing.

6.4.1 Impulse Response

The impulse response of a network is found, again, by using the general transfer function equation

$$Y(s) = X(s)T(s)$$

If the transfer function is defined as the output voltage V_O divided by the input voltage V_I, and the excitation is the unit impulse function, the output is the s-plane representation of the impulse response. In the s-plane, the impulse function is the constant 1. The response is often identified as $G(s)$. Substituting these into the preceding equation, we get

$$G(s) = 1[T(s)] = T(s) \tag{6-7}$$

The transform of the impulse response of a network is the network transfer function. The impulse response itself is the time domain function

$$g(t) = \mathscr{L}^{-1}[G(s)] \tag{6-8}$$

We can gain some familiarity with impulse responses by finding the responses for a first-order and a second-order circuit.

EXAMPLE
6–8

Find the impulse response for the RC circuit shown in Figure 6–20(a).

The Laplace transform of the impulse response was defined as the voltage transfer function. That is,

$$G(s) = T(s) = \frac{V_2(s)}{V_1(s)}$$

First, we need to transform the circuit. The transformed circuit is shown in Figure 6–20(b). From the voltage divider rule, we obtain

$$V_2(s) = V_1(s)\frac{Z_C(s)}{Z_T(s)}$$

where $Z_T(s)$ is the sum of both impedances. Solving for the transfer function, we get

$$\frac{V_2(s)}{V_1(s)} = G(s) = \frac{Z_C(s)}{Z_T(s)} = \frac{1/6s}{5 + (1/6s)} = \frac{0.03333}{s + 0.03333}$$

This is the transform of the impulse response. To find the response itself, we use transform pair 2 of Table 4–1 and operation pair A of Table 4–2 to get

$$g(t) = 0.03333e^{-0.03333t} \text{ V}$$

The curve of this equation is plotted in Figure 6–21.

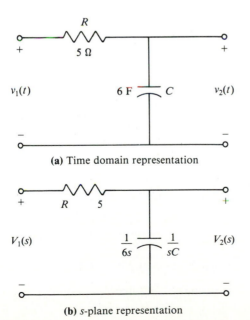

(a) Time domain representation

(b) s-plane representation

FIGURE 6–20 RC circuit for Example 6–8

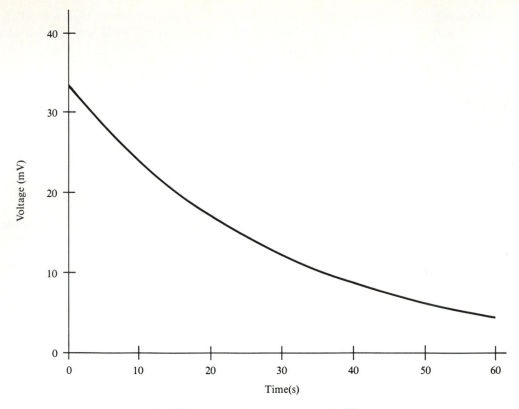

FIGURE 6–21   Impulse response for RC circuit

EXAMPLE
6–9   Find the impulse response in the time domain for the series RLC circuit shown in Figure 6–22(a).

The transformed circuit is shown in Figure 6–22(b). In this series circuit, we can again use the voltage divider rule. The transfer function is then

$$T(s) = G(s) = \frac{V_2(s)}{V_1(s)} = \frac{Z_C(s)}{Z_T(s)} = \frac{1/6s}{10s + 5 + (1/6s)}$$

$$= \frac{0.01667}{s^2 + 0.5s + 0.01667}$$

To put the denominator in factored form, we use the formula for finding the roots of quadratics. We have:

$$s_1, s_2 = -b/2 \pm \sqrt{(b/2)^2 - c}$$

$$= -0.25 \pm \sqrt{(0.25)^2 - 0.01667}$$

$$= -0.03590, -0.4641$$

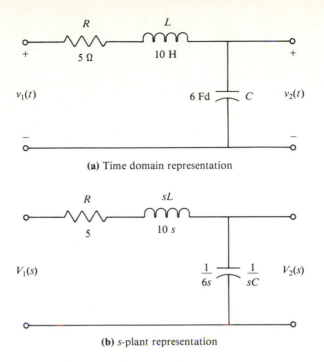

(a) Time domain representation

(b) s-plant representation

FIGURE 6–22 RLC circuit for Example 6–9

The transfer function is

$$T(s) = G(s) = \frac{0.01667}{(s + 0.03590)(s + 0.4641)}$$

Using partial fraction expansion, we get

$$G(s) = \frac{0.03893}{s + 0.03590} - \frac{0.03893}{s + 0.4641}$$

From pair 2 of Table 4–1 and pair A of Table 4–2, we obtain the impulse response:

$$g(t) = 0.03893\epsilon^{-0.03590t} - 0.03893\epsilon^{-0.4641t} \text{ V}$$

The curve of this equation is illustrated in Figure 6–23.

The impulse responses plotted in Figures 6–21 and 6–23 are similar in their general shape, particularly in the region where the second term in the response for the RLC circuit becomes very small as t increases. In that region, the other term is much larger in amplitude and is therefore called the *dominant term*. If the time constants had been closer in value, the curves would have differed more.

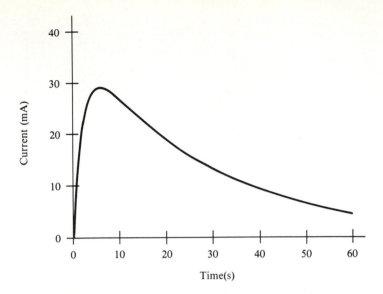

FIGURE 6–23 Impulse response for RLC circuit

The impulse response is of considerable interest partly because it is equal to the Laplace transform of the network transfer function. Also, its spectrum has constant amplitude across the frequency band, so the output should provide information about system operation over a wide frequency band. However, in a restricted frequency band, such as an amplifier passband, the total energy in an impulse response will be of low amplitude. The actual output level will be very low for a true impulse. A very narrow pulse used to simulate the impulse would provide very low-level output, also.

Because of this low amplitude, the impulse response is little used in circuit testing. It is more often used in computation, particularly since tables of voltage transfer functions of circuits are generally available.

6.4.2 Step Response

The step function does not have the flat frequency spectrum that the impulse function does. Rather, its spectral amplitude is inversely proportional to frequency. When the step function is used to test a system, low-frequency response problems may be more obvious than high-frequency response problems.

A step function may be generated by a fast switching circuit, which is more easily produced than a fast impulse driver. Also, because the step function is not inherently a low-energy signal, reasonable circuit output levels can be expected when it is used in testing. Therefore, the step function is a good choice of waveform for testing circuits.

The step response is often denoted by $H(s)$, and the Laplace transform of the unit step function is $1/s$. Substituting these into Equation (6–1), we get, for the Laplace transform of the step response,

$$H(s) = \frac{T(s)}{s} \qquad (6\text{–}9)$$

The step response in the time domain is

$$h(t) = \mathscr{L}^{-1}[H(s)] \qquad (6\text{–}10)$$

Equation (6–9) shows that the transform of the step response is the transfer function divided by s. This adds some complexity to the process of finding the step response over and above that involved in finding the impulse response.

EXAMPLE Find the step response of the RC circuit in Figure 6–20.
6–10 This is the same circuit for which we found the impulse response in Example 6–8. In that example, after transforming the circuit, we found the transfer function to be

$$T(s) = \frac{0.03333}{s + 0.03333}$$

The output will be

$$V_2(s) = V_1(s)T(s)$$

where the input is the unit step function, $1/s$, in the s-plane. The step response in the s-plane is then

$$H(s) = V_2(s) = \frac{0.03333}{s(s + 0.03333)}$$

Expanding this, we obtain

$$H(s) = \frac{1}{s} - \frac{1}{s + 0.03333}$$

Using transform pair 2 of Table 4–1 and operation pair A of Table 4–2, we get the step response

$$h(t) = 1 - \epsilon^{-0.03333t} \text{ V}$$

The step response for the RC circuit is shown in Figure 6–24.

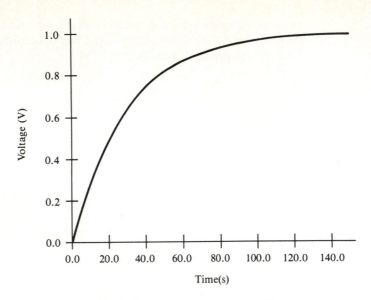

FIGURE 6–24 Step response for RC circuit

EXAMPLE 6–11 Find the step response for the RLC circuit in Example 6–9.

The circuit is illustrated in Figure 6–22. In Example 6–9, its transfer function was found to be

$$T(s) = \frac{0.01667}{(s + 0.03590)(s + 0.4641)}$$

We can use Equation (6–9) to get the transform of the step response:

$$H(s) = \frac{T(s)}{s} = \frac{0.01667}{s(s + 0.03590)(s + 0.4641)}$$

After expansion, we have

$$H(s) = \frac{1.001}{s} - \frac{1.084}{s + 0.03590} + \frac{0.08388}{s + 0.4641}$$

Transforming this with transform pairs 1 and 2 of Table 4–1 and operation pair A of Table 4–2 gives

$$h(t) = 1.001 - 1.084\epsilon^{-0.03591t} + 0.08388\epsilon^{-0.4641t} \text{ V}$$

The step response for the RLC circuit is plotted in Figure 6–25. As with the impulse responses, the step responses are quite similar in shape due to the presence of a clearly dominant pole.

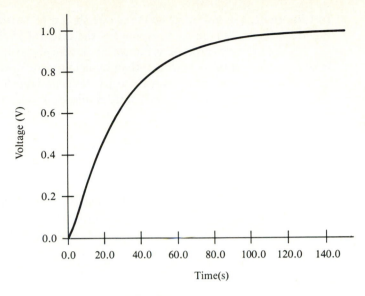

FIGURE 6–25   Step response for RLC circuit

6.4.3   Comparison of Step and Impulse Responses

Let us compare the impulse and step responses for the RC and the RLC circuits. For the RC circuit,

$$g(t) = 0.03333\epsilon^{-0.03333t} \text{ V}$$

and

$$h(t) = 1 - \epsilon^{-0.03333t} \text{ V}$$

The responses for the RLC circuit are

$$g(t) = 0.03893\epsilon^{-0.03590t} - 0.03893\epsilon^{-0.4641t} \text{ V}$$

and

$$h(t) = 1.001 - 1.084\epsilon^{-0.03591t} + 0.08388\epsilon^{-0.4691t} \text{ V}$$

The magnitudes of the terms for the impulse responses are much smaller than those for the step responses in both cases. Also, finding the step responses was more difficult because of the presence of an extra term. This in turn was due to the more complex s-plane expression for the step function.

Using operation pair E from Table 4–2, we obtain

$$h(t) = \int_{0}^{t} g(t) \, dt \qquad (6\text{–}11)$$

In Chapter 2, the unit step function was shown to be the integral of the impulse function. Similarly, the step response is the integral of the impulse response.

If the impulse response is known, it may be easier to find the step response by integrating the impulse response than by using transform analysis. The availability of tables of transfer functions of electric circuits and mechanical and electromechanical devices in the literature is of considerable assistance in finding impulse and step responses.

6.5   BODE DIAGRAMS

Earlier in this chapter, we used Laplace transform concepts to develop the transfer functions of networks. In the previous section, we said that if the transfer functions are evaluated over the complete s-plane, the responses to the entire range of exponential inputs can be found.

The steady-state frequency responses for various sinusoidal inputs are probably the most important responses for electronic systems in general. They are used in evaluating the performance of many types of equipment, even those for which a single-frequency sine wave is not a typical signal.

The frequency response may be determined by evaluating the transfer function along the $j\omega$, or frequency, axis. The general form of the transfer function is in Equation (6–4),

$$T(s) = \frac{N(s)}{D(s)}$$

where $N(s)$ and $D(s)$ are polynomials in s. These polynomials can be factored into a number of simpler terms that make it much easier to evaluate the response of the system. An nth-degree polynomial will have n first-degree factors. All first-degree terms with real roots should be factored out; quadratic terms that have complex roots need not be factored further. The result of these operations will be the transfer function in the form

$$T(s) = \frac{K(\tau_1 s + 1) \cdots (s^2/\omega_{n1}^2 + 2\zeta s/\omega_{n1} + 1) \cdots}{s^n (\tau_2 s + 1) \cdots (s^2/\omega_{n2}^2 + 2\zeta s/\omega_{n2} + 1) \cdots} \qquad (6\text{--}12)$$

where n is a positive or negative integer.

When finding the frequency response, we let s equal $j\omega$ to get the steady-state transfer function $T(j\omega)$. Generally, the terms are complex, so $T(j\omega)$ is a vector whose magnitude is the product of the magnitudes of the numerator terms divided by the product of the magnitudes of the denominator terms; that is,

$$|T(j\omega)| = \frac{K|j\tau_1 \omega + 1| \cdots |-\omega^2/\omega_{n1}^2 + j2\zeta\omega/\omega_{n1} + 1| \cdots}{|j\omega^n||j\tau_2 \omega + 1| \cdots |-\omega^2/\omega_{n2}^2 + j2\zeta\omega/\omega_{n2} + 1| \cdots}$$

The magnitudes of the individual terms can be converted to decibels using the equation

$$|T(j\omega)|_{dB} = 20 \log_{10} |T(j\omega)|$$

The magnitude in decibels of the complete expression $T(j\omega)$ can then be found by adding the decibel magnitudes of the individual numerator factors and subtracting the magnitudes of the denominator factors.

The phase angle of $T(j\omega)$ is the sum of the phase angles of the numerator terms minus the phase angles of the denominator terms. If the amplitude and phase responses of the individual factors are plotted, the complete responses can be found graphically.

The responses are usually plotted on semilogarithmic, or semilog, graph paper. This kind of graph paper is illustrated in Figure 6–26. The abscissa, along which frequency is plotted, is the logarithmic axis. There will often be three or more cycles along the frequency axis, each cycle representing a decade of frequency. The paper shown in the figure has only two cycles and thus can cover a frequency range of $1:100$. For instance, the paper could be used for a range of 0.1 to 10 in frequency. Frequency can be expressed in radians per second or hertz. To simplify plotting, we will express frequency in radians per second to start with.

In the amplitude plot, the ordinate is the amplitude ratio in decibels (dB). The decibel is a logarithmic ratio, so the amplitude vs. frequency plot is really a log-log plot. The phase angle is plotted as the ordinate of the phase response, usually in degrees. This makes the process of summing up the individual components due to each term fairly simple.

When circuits are cascaded, if loading effects are either negligible or accounted for, the total gain in dB is the sum of the individual circuit gains. Similarly, the total phase shift is the sum of the phase shifts of the individual circuits. Plots of the sinusoidal responses of complex systems can be generated by summing the dB amplitude and phase shift factors for each factor in the transfer functions. These plots of the logarithm of amplitude and phase shift versus the logarithm of frequency are called *Bode plots* or *Bode diagrams,* in acknowledgement of pioneering work done in this area by H. W. Bode.

6.5.1 Amplitude and Phase Responses of Simple Factors

The responses of simple factors can be plotted approximately using only a straight-edge. Any errors incurred will be small, particularly for the amplitude responses. As a result, the straight-line approximations are often used for initial estimates of circuit performance.

6.5.1.1 Constant Gain, K The first step in evaluating the gain constant K is to divide each numerator and denominator factor having a constant term by the number that makes the constant term equal to 1. K will then be the original constant gain term multiplied by the numerator constants and divided by the denominator constants.

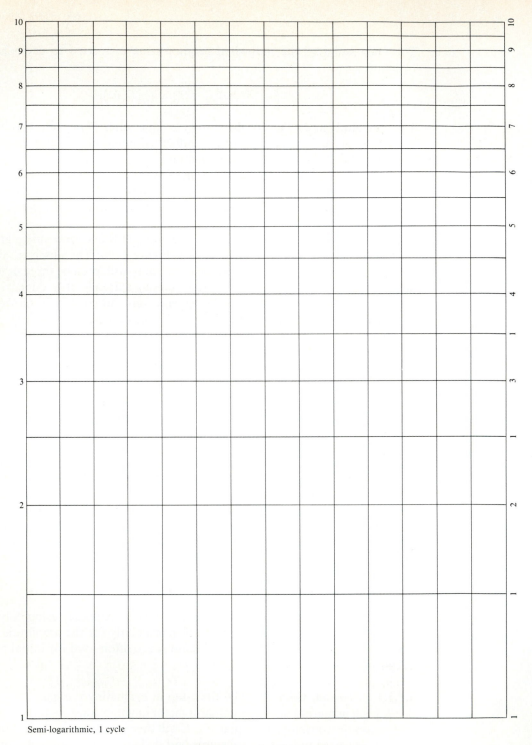

Semi-logarithmic, 1 cycle

FIGURE 6–26 Semilog graph paper

For example, if the factored polynomial is

$$T(s) = \frac{10(s^2 + 3s + 2)}{(s + 1)(s + 2)(s + 5)} \qquad (6\text{–}13)$$

then the complete multiplier is $2/(2)(5)$, or 0.2, and

$$T(s) = \frac{2(0.5s^2 + 1.5s + 1)}{(s + 1)(0.5s + 1)(0.2s + 1)}$$

and $K = 2$. As mentioned earlier, magnitudes are expressed in decibels, so

$$K_{dB} = 20 \log_{10} K$$

The constant gain factor causes no phase shift; its only effect on the transfer function is to provide a constant gain. If the transfer function has no s^n factors in the numerator or denominator, then K is the DC gain of the system. To illustrate, we can find the DC gain of the circuit represented by Equation (6–13) by evaluating it with $s = j\omega = 0$:

$$T(0) = \frac{10(2)}{1(2)(5)} = 2 = 6 \text{ dB}$$

This is plotted in Figure 6–27. Note that the amplitude response is a straight, horizontal line; there is no phase shift, since K is a positive scalar.

6.5.1.2 Pole at the Origin A factor whose transfer function is

$$T(s) = \frac{1}{s}$$

has a pole at the origin.

To find the amplitude and phase response for such a factor, we substitute $j\omega$ for s, giving

$$T(j\omega) = \frac{1}{j\omega}$$

where j in the denominator indicates a phase shift of $-\pi/2$ radians, or $-90°$, at all frequencies. The amplitude is inversely proportional to ω, the frequency in radians; that is,

$$|T(j\omega)| = \frac{1}{\omega} \qquad (6\text{–}14)$$

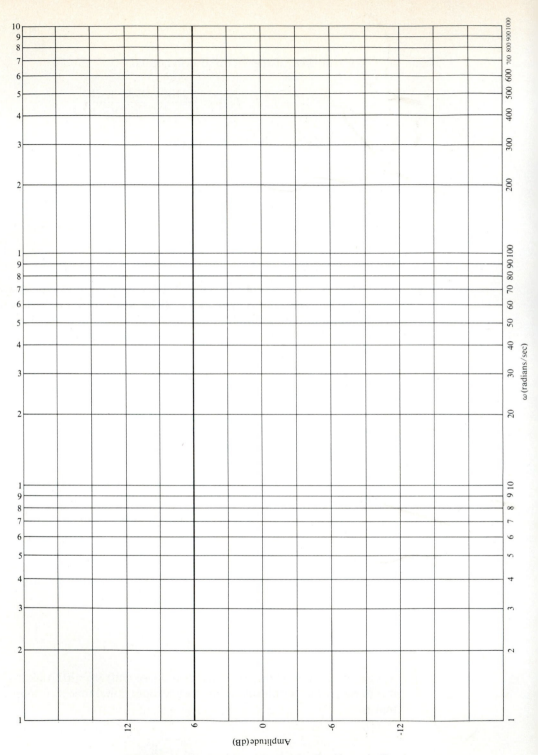

FIGURE 6–27 Bode amplitude plot for $K_{dB} = 6$ dB

In decibels, this is

$$|T(j\omega)|_{dB} = 20\log_{10}\left(\frac{1}{\omega}\right) = -20\log_{10}\omega \qquad (6\text{--}15)$$

The phase angle is given by

$$\angle T(j\omega) = \tan^{-1}\left(\frac{1}{j1}\right) = -90° = -\frac{\pi}{2}\text{ radians} \qquad (6\text{--}16)$$

The amplitude and phase responses are plotted on semilog paper in Figure 6–28. On this paper, both curves are straight lines. If we substitute $\omega = 1$ into Equation (6–14), the amplitude response is 0 dB at a frequency of 1 radian per second. The amplitude decreases at a rate of 20 dB for a frequency ratio of 10 to 1, usually expressed as -20 dB per decade of frequency. This rate of decrease is equal to 6.021 dB per octave of frequency, usually rounded off to 6 dB per octave.

The phase shift is a horizontal line at $-90°$. The amplitude and phase responses can be plotted easily with a straightedge on semilog paper if we express amplitude in dB and frequency in radians per second. The response shown is that of an integrator, a result not unexpected, from inspection of the general transfer function $1/s$. Table 4–2 shows that this response is the Laplace transform of an integral.

6.5.1.3   Zero at the Origin   The response caused by a zero at the origin of the s-plane is

$$T(s) = s$$

The steady-state transfer function of this response is

$$T(j\omega) = j\omega$$

This function has a magnitude of

$$|T(j\omega)| = \omega \qquad (6\text{--}17)$$

or, in dB,

$$|T(j\omega)|_{dB} = 20\log_{10}\omega \qquad (6\text{--}18)$$

When $\omega = 1, \log(\omega) = 0$, so the magnitude is 0 dB. When $\omega = 10$, $\log(\omega) = 1$, so the magnitude is 20 dB. At $\omega = 0.1$, the magnitude becomes -20 dB. For each tenfold increase in frequency, the magnitude increases 10 times, a relationship usually expressed as 20 dB increase per decade of frequency. Doubling the frequency (i.e., increasing it by an octave) will provide an increase of $20\log_{10}2$, or 6.021 dB, which is usually rounded off to 6 dB increase per octave.

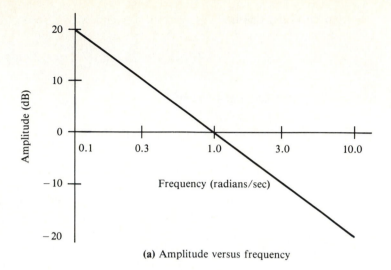

(a) Amplitude versus frequency

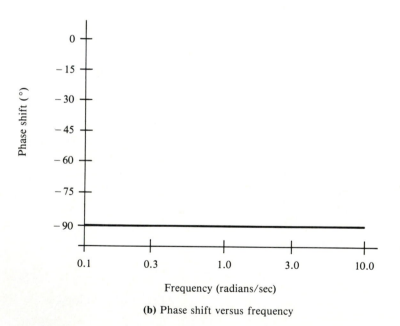

(b) Phase shift versus frequency

FIGURE 6–28 Pole at origin

The amplitude response of a circuit with a zero at the origin is illustrated in Figure 6–29(a). The phase shift, which is found by

$$\angle T(j\omega) = 90° = \frac{\pi}{2} \text{ radians} \tag{6–19}$$

is illustrated in Figure 6–29(b).

The effect of a zero at the origin on the response is a 20-dB increase per decade of frequency with a magnitude of 0 dB at $\omega = 1$ radian per second. It will also add a constant phase shift of 90° to that of any other terms. In Chapter 4, we showed that a differentiator has this type of response.

6.5.1.4 Multiple Poles or Zeros at the Origin When there are several s-terms in either the denominator or the numerator of the transfer function, each will provide its own amplitude or phase change in the steady-state transfer function $T(j\omega)$. These effects can be added together to get the total response. A pair of poles at the origin indicates -40 dB per decade, twice the slope in the amplitude response indicated by a single pole. The phase response will also be doubled, to $-180°$.

The amplitude and phase responses for multiple poles or zeros at the origin are

$$|T(j\omega)| = \omega^n \tag{6–20}$$

or, in decibels,

$$|T(j\omega)|_{dB} = 20n \log_{10} \omega \tag{6–21}$$

and

$$\angle T(j\omega) = \frac{n\pi}{2} \tag{6–22}$$

In these equations, n is the power of the s^n term when this term is in the numerator of the transfer function. Plots of the response terms where $n = \pm 1$ and $n = \pm 2$ are shown in Figure 6–30. The magnitude of n is equal to the number of integrators or differentiators in the circuit.

6.5.1.5 Pole on the Real Axis The general transfer function term for a pole on the real axis is

$$T(s) = \frac{1}{\tau s + 1}$$

Substituting $j\omega$ for s in this equation, we get

$$T(j\omega) = \frac{1}{j\omega\tau + 1} \tag{6–23}$$

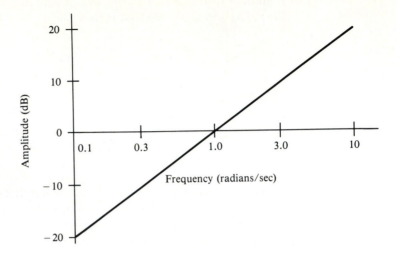

(a) Amplitude versus frequency

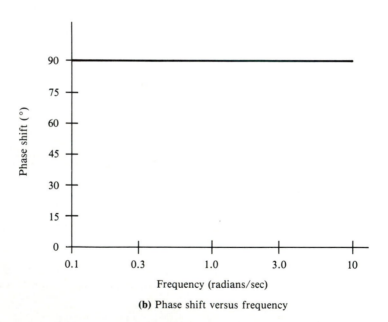

(b) Phase shift versus frequency

FIGURE 6–29 Zero at origin

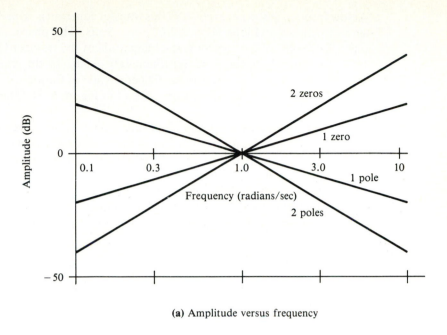

(a) Amplitude versus frequency

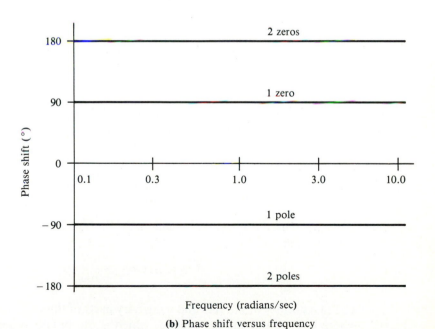

(b) Phase shift versus frequency

FIGURE 6–30 Multiple poles and zeros at origin

At low frequencies, where $\omega\tau \ll 1$, this transfer function is approximately $1\angle 0°$. For $\omega\tau = 1$, $|T(j\omega)| = 1/|1 + j1| = 0.7071$, or -3 dB. The phase angle at this point is $\tan^{-1} 1$, or $-45°$. When $\omega\tau \gg 1$, the magnitude of the transfer function approaches $1/\omega\tau$, and the phase angle is approximately $-90°$. In this frequency range, the magnitude decreases at a rate of 20 dB per decade of frequency. The amplitude and phase for a pole on the real axis are plotted in Figure 6–31. The response is that of a simple low-pass filter.

An approximation to the amplitude response may be made by drawing a horizontal line at 0 dB for frequencies up to $\omega = 1/\tau$ radians per second and a second line intersecting the first $\omega = 1/\tau$ radians/second, with a slope of -20 dB per decade. These are also plotted in Figure 6–31(a). The frequency of the intersection of the straight lines is called the *break,* or *corner, frequency.* The exact curve is asymptotic to the approximate curve. The break frequency is the point where the real and imaginary components in the denominator of Equation (6–23) are equal in magnitude, that is, where $\omega = 1/\tau$.

The maximum error of the amplitude approximation, at $\omega = 1/\tau$, is 3 dB. At $\omega = 0.5/\tau$ and $\omega = 2/\tau$, the errors are both 1 dB. These errors can be used to give a more accurate plot. In many cases, however, the straight-line plot is accurate enough for design work.

An approximation is sometimes used for the phase response. This is shown in Figure 6–31(b). It is constant from 0° up to one-tenth of the break frequency, that is, $\omega = 0.1/\tau$, and then it is a straight line sloping to $-90°$ at 10 times the break frequency, that is, $\omega = 10/\tau$. At the break frequency $\omega = 1/\tau$, the phase shift is $-45°$.

Table 6–1 lists values from the Bode plots for negative real-axis poles, the straight-line approximations of these values, and the errors involved in using the approximations. Because the phase shift errors may cause serious problems in design calculations, the phase shift approximations should be used only in the preliminary stages of design.

EXAMPLE 6–12   Find the breakpoint of the transfer function $T(s) = 5/(s + 10)$. Draw the approximate and exact Bode plots for this function.

The steady-state transfer function is

$$T(j\omega) = \frac{5}{j\omega + 10}$$

Dividing numerator and denominator by 10 to put the denominator in the proper form, we get

$$T(j\omega) = \frac{0.5}{1 + j(\omega/10)} \tag{6–24}$$

The magnitudes of the real and imaginary parts of the complex term are thus equal where $\omega = 10$ radians per second, which is the break frequency. The break frequency may also be expressed in hertz as $f = \omega/2\pi = 10/2\pi$, or 1.592 Hz, here.

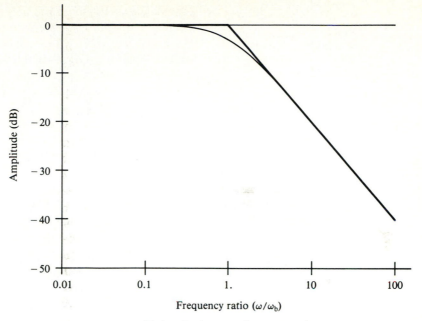

(a) Amplitude versus frequency ratio

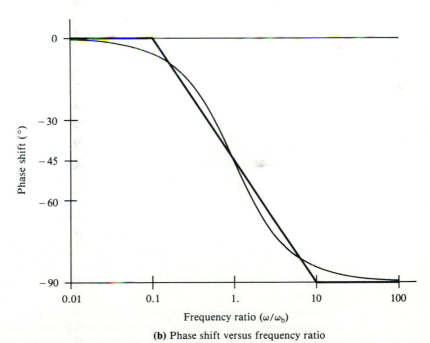

(b) Phase shift versus frequency ratio

FIGURE 6–31 First-order pole on negative real axis

TABLE 6–1
Amplitude and
Phase Shift for
a Negative
Real-axis Pole,
Exact and
Approximate

$\omega\tau$	0.10	0.50	0.76	1	1.31	2	5	10
Exact amplitude, dB	−0.04	−1.0	−2.0	−3.0	−4.3	−7.0	−14.2	−20.04
Amplitude approximation, dB	0	0	0	0	−2.3	−6.0	−14.0	−20.0
Error, dB	0.04	1.0	2.0	3.0	2.0	1.0	0.2	0.04
Exact phase shift, degrees	−5.7	−26.6	−37.4	−45	−52.7	−63.4	−78.7	−84.3
Phase shift approximation, degrees	0	−31.5	−39.5	−45	−50.3	−58.5	−76.5	−90.0
Error, degrees	5.7	−4.9	−2.1	0	2.4	4.9	2.2	−5.7

The amplitude approximation is 0 dB for frequencies up to the break frequency and drops at a slope of −20 dB per octave above the break frequency. The exact amplitude response is 1 dB below the approximate response at one-half and twice the break frequency. Both of the amplitude curves are plotted in Figure 6–32(a).

The phase response approximation is 0° at $\omega = 1$ radian per second and slopes to −90° at $\omega = 100$ radians per second. Both the approximation and the exact phase response are shown in Figure 6–32(b).

6.5.1.6 Zero on the Real Axis The transfer function for a zero on the real axis will be the reciprocal of that for a pole at the same location, that is,

$$T(s) = s + 1$$

The amplitude response in dB and the phase response will therefore be the negative of those for a real-axis pole.

When we substitute $j\omega$ for s in the transfer function, we get

$$T(j\omega) = j\omega\tau + 1 \tag{6–25}$$

The amplitude of this function at frequencies well below the break frequency will be constant at 0 dB. At frequencies much above the break point, the amplitude will increase at a rate of 20 dB per decade. This is illustrated in Figure 6–33(a). The low-frequency phase shift will be 0°. At the break frequency, the phase shift will be 45°. When $\omega \gg 1/\tau$, the phase shift approaches 90°. The phase shift is plotted in Figure 6–33(b).

The straight-line amplitude approximation is 0 dB up to the break frequency, and it increases at a rate of 20 dB per octave above that point. The errors can be seen to be of the same magnitude as those for a pole on the real axis. At $\omega = 0.5/\tau$ and $2/\tau$, the approximation is low by 1 dB, and at the break frequency the straight-

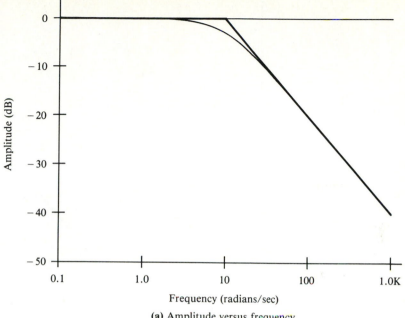

(a) Amplitude versus frequency

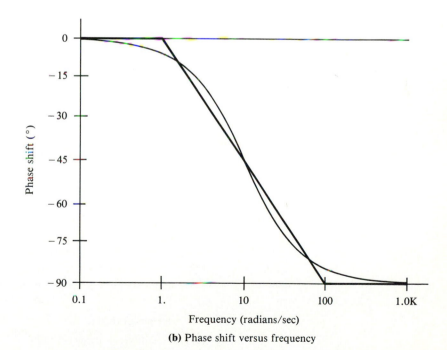

(b) Phase shift versus frequency

FIGURE 6–32 Bode plots for $T(s) = 0.5/[1 + j(0.1\omega)]$

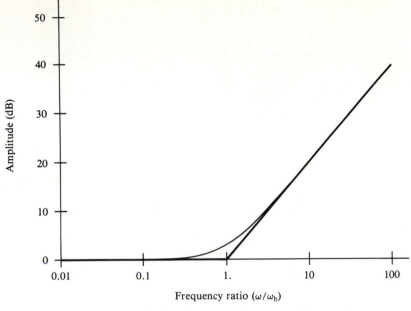

(a) Amplitude versus frequency ratio

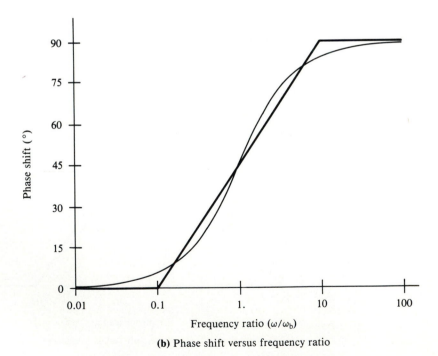

(b) Phase shift versus frequency ratio

FIGURE 6–33 First-order zero on negative real axis

line plot is low by 3 dB. The phase shift approximation is $0°$ up to $\omega = 0.1/\tau$ and $90°$ at and above $\omega = 10/\tau$. This is the negative of the approximation for the first-order pole on the real axis. Both the exact and the approximate decibel amplitude and phase responses for the real zero are the inverse of the response for the real pole.

6.5.1.7   Complex Pole   The type of transfer function factor that has complex conjugate poles is

$$F_1(s) = \frac{1}{(s^2/\omega_n^2) + (2\zeta s/\omega_n) + 1}$$

When $\zeta \geq 1$, the quadratic may be factored into two real poles. If $\zeta < 1$, the factors are complex and the response is more complicated. Substituting $j\omega$ for s in the transfer function, we get the steady-state transfer function

$$F_1(j\omega) = \frac{1}{(j\omega/\omega_n)^2 + (j2\zeta\omega/\omega_n) + 1} = \frac{1}{1 - (\omega^2/\omega_n^2) + (j2\zeta\omega/\omega_n)} \quad (6\text{–}26)$$

When $\omega \ll \omega_n$, the magnitude of $F_1(j\omega)$ approaches 1, or 0 dB. At frequencies well above ω_n, the magnitude decreases at -40 dB per decade of frequency, twice the drop due to a single pole. The response near ω_n, the natural frequency of the system, depends very much on the value of the imaginary term, $j2\zeta\omega/\omega_n$, so the general response cannot be described in any simple fashion. This is true of both the amplitude and phase responses.

The amplitude is

$$|F_1(j\omega)| = \frac{1}{[(1 - \omega^2/\omega_n^2)^2 + 4(\zeta\omega/\omega_n)^2]^{1/2}} \quad (6\text{–}27)$$

The phase response is given by

$$\angle F_1(j\omega) = -\tan^{-1}\frac{2\zeta\omega/\omega_n}{1 - (\omega^2/\omega_n^2)} \quad (6\text{–}28)$$

These are shown in Figure 6–34 as families of curves for values of the damping ratio ζ between 0.05 and 1. The amplitude has a peak in the neighborhood of the natural frequency ω_n for small values of the parameter ζ. The amplitude at frequencies well below ω_n approaches 0 dB, and well above the natural frequency it decreases at a rate of 40 dB per decade, twice as fast as with a simple real pole.

If ζ equals 1, the denominator can be factored into two real equal roots, as discussed earlier. Summing the amplitude factors, in decibels, for the poles should give a response twice that found for a single real pole in Section 6.5.1.5. Inspection of the curve in Figure 6–32(a) and the curve for ζ equal to 1 in Figure 6–34(a) shows this to be true. Similarly, the phase response in Figure 6–34(b) is twice that for a single real pole in Figure 6–32(b).

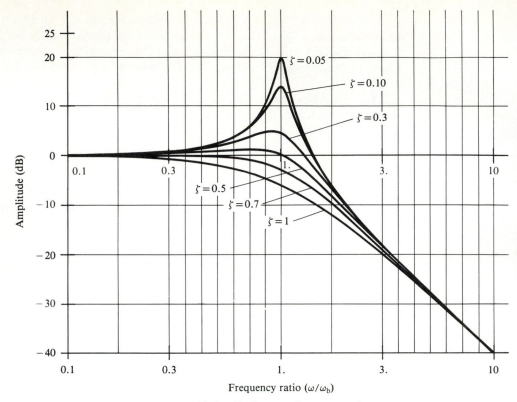

(a) Amplitude versus frequency ratio

FIGURE 6–34   *Complex poles*

Responses due to complex pole pairs are difficult to approximate using straight lines. In evaluating complex transfer functions that include complex pole pairs, it is possible to plot the sum of the complete approximations for any first-order terms. Then we can add to them the phase shift and magnitude in dB from the curves in Figure 6–34 to get an approximation for the complete response.

6.5.1.8   Complex Zero   The transfer function that has complex zero roots is

$$T(s) = \left(\frac{s}{\omega_n}\right)^2 + \frac{2\zeta s}{\omega_n} + 1$$

The steady-state transfer function is found by letting $s = j\omega$, which gives

$$T(j\omega) = (j\omega/\omega_n)^2 + \frac{j2\zeta\omega}{\omega_n} + 1$$

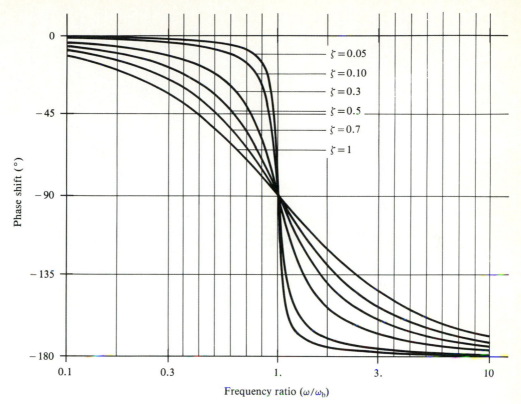

(b) Phase shift versus frequency ratio

FIGURE 6–34 Continued

Collecting real and imaginary terms, we get

$$T(j\omega) = 1 - \left(\frac{\omega}{\omega_n}\right)^2 + \frac{j2\zeta\omega}{\omega_n} \tag{6–29}$$

This has an amplitude response of

$$|T(j\omega)| = \left[\left(\frac{1 - \omega^2}{\omega_n^2}\right)^2 + \frac{4\zeta^2\omega^2}{\omega_n^2}\right]^{1/2} \tag{6–30}$$

the reciprocal of that for the complex pole. Therefore, it should approach 0 dB at low frequencies and increase at approximately 40 dB per decade at high frequencies. The responses will also depend on the magnitude of the damping ratio ζ. This is plotted as a family of curves in Figure 6–35(a) for values of ζ between 0.05 and 1.

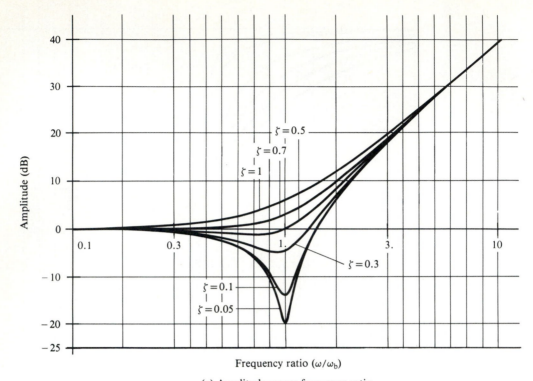

(a) Amplitude versus frequency ratio

FIGURE 6–35   Complex zeros

These curves are the negative of the amplitude curves, in decibels, for a pair of complex poles. The phase response is

$$\angle T(j\omega) = \tan^{-1}\frac{2\omega/\omega_n}{1 - (\omega/\omega_n)^2} \qquad (6\text{–}31)$$

the negative of the phase response for a pair of complex poles.

For $\omega \ll \omega_n$, the phase shift is close to $0°$. When $\omega = \omega_n$, the phase shift is $\tan^{-1}\infty$, or $90°$. For large values of ω, the tangent approaches zero from a negative value, so the angle approaches $180°$. This phase response is the negative of that found for a pair of complex poles. A family of curves for values of ζ from 0.05 to 1 is plotted in Figure 6–35(b).

As with complex poles, the responses of complex zeros are not easily approximated. The curves for both decibel amplitude and phase are the negative of those for complex poles, so we could get the data for complex zeros and complex poles from any one such set of curves.

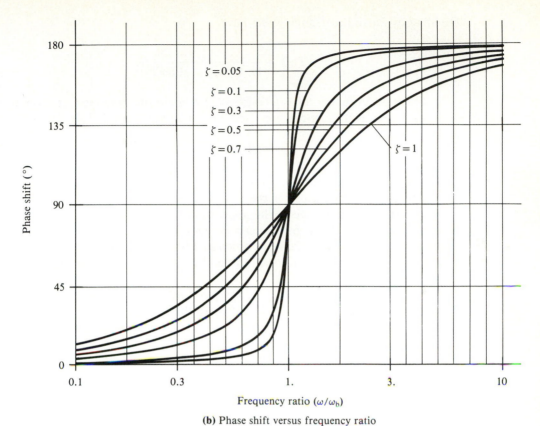

(b) Phase shift versus frequency ratio

FIGURE 6–35 *Continued*

6.5.2 System Bode Plots

The evaluation of a comparatively complex transfer function consisting of several first-order poles and zeros can be done fairly easily. The tools required are semilog graph paper, a straightedge, and the foregoing techniques.

EXAMPLE 6–13

A comparatively simple system has the transfer function

$$T(s) = \frac{30}{s^2 + 21s + 20}$$

Draw the approximate Bode plots for the system. Then, using these plots and the error data listed in Table 6–1, sketch the exact curves.

The quadratic expression in the denominator can be factored to find its roots:

$$T(s) = \frac{30}{(s + 1)(s + 20)}$$

Dividing by 20 and rearranging, we have

$$T(s) = \frac{1.5}{(1 + j\omega)[1 + (j\omega/20)]}$$

The transfer function has a constant gain factor of 1.5 times and two real-axis poles whose break frequencies are 1 and 20 radians per second. The gain is equal to

$$K_{dB} = 20 \log 1.5 = 3.5 \text{ dB}$$

and there is no phase shift due to this term.

The break frequencies for the poles are well separated, so their individual effects should be readily seen. The approximate curves will each have a constant amplitude level of 0 dB at frequencies below their break frequencies of 1 and 20 radians per second. Then, above its individual break frequency, each will decrease at a rate of −20 dB per decade. All three of these approximate amplitudes curves are plotted in Figure 6–36(a).

From Table 6–1, for each pole factor, the exact amplitude curves are 1 dB below the approximate curves at one-half and twice the break frequencies. At the break

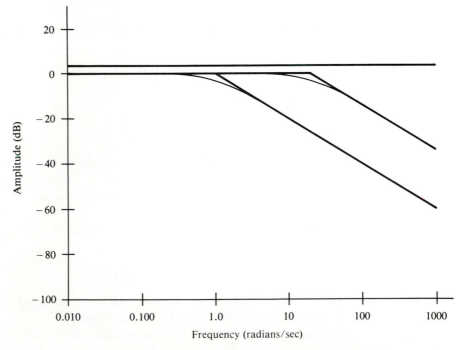

(a) Approximate and exact amplitude-versus-frequency plot of individual components

FIGURE 6–36   Continued

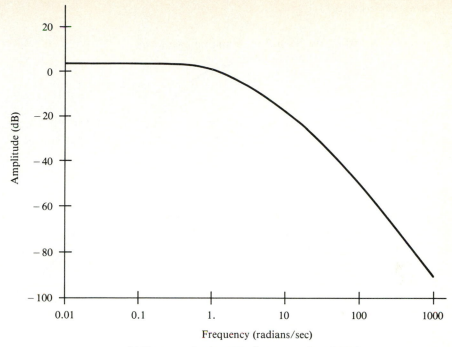

(b) Exact amplitude-versus-frequency plot of total system

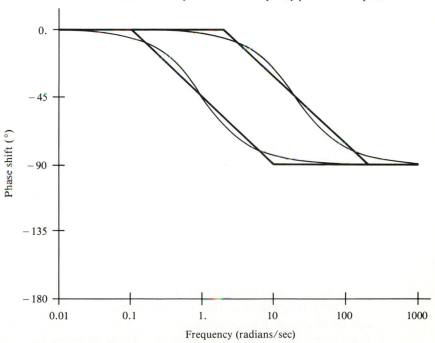

(c) Plot of approximate and exact phase shift versus frequency for individual components

FIGURE 6–36 Continued

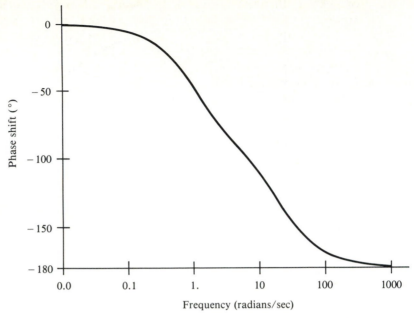

(d) Plot of exact phase shift versus frequency for total system

FIGURE 6–36 *Simple system*

frequencies, the exact curves are 3 dB below the approximate curves. Well away from the break frequencies, the exact and approximate responses are equal.

The exact amplitude curves for the total system are shown in Figure 6–36(b). As discussed in Chapter 2, when adding curves, at each value of the independent variable (frequency) both the magnitude and the slope of the sum are the sum of the magnitudes and slopes of each individual component.

The approximate response is readily found for this system. At all frequencies the gain factor is 3.5 dB. The two poles have 0-dB magnitudes at low frequencies, so the total magnitude from DC to the first break frequency, 1 radian per second, is 3.5 dB. From that point to the second break frequency, 20 radians per second, the amplitude drops at -20 dB per decade, due to the factor $1 + j\omega$. Above 20 radians per second, the second pole causes a further drop of -20 dB per decade. The total drop in this region of the response is -40 dB per decade.

Since the break points are separated in frequency, the exact curves can be found by subtracting the errors given in Table 6–1 from the approximate curves.

Both denominator factors represent real-axis poles, so their phase responses will be 0° at very low frequencies and will become $-90°$ at very high frequencies. Each factor produces a phase shift of $-45°$ at its break frequency, 1 or 20 radians per second. The approximations will be 0° up to 0.1 times the break frequency (0.1 or 2 radians per second), with a slope of $-45°$ per decade, and will go up to a value of $-90°$ at 10 times the break frequency (10 or 200 radians per second). The approximate and exact phase shifts due to each factor are plotted in Figure 6–36(c).

As with the amplitude responses, the total phase shift is the sum of the individual phase shifts. The total phase shift will be $0°$ at very low frequencies and will approach a maximum of $180°$ at much higher frequencies. The exact phase shift for the complete system are shown in Figure 6–36(d). The maximum errors of $10.6°$ and $-10.6°$ occur at 2 and 10 radians per second, respectively.

EXAMPLE 6–14

Draw the approximate Bode plots for a system with transfer function

$$T(s) = \frac{10s}{4s^2 + 88s + 160}$$

We can divide numerator and denominator by 160 to make the constant terms in all factors equal to 1. Then, factoring the denominator, we obtain the transfer function in the form that best lends itself to use in developing Bode plots, that is,

$$T(s) = \frac{0.0625s}{[1 + (s/2)][1 + (s/20)]}$$

This transfer function has a constant gain factor, a zero at the origin (representing a differentiator), and two poles on the real axis. Letting $s = j\omega$, we get

$$T(j\omega) = \frac{0.0625(j\omega)}{[1 + (j\omega/2)][1 + (j\omega/20)]}$$

The break frequencies for the poles are at $\omega = 2$ and $\omega = 20$ radians per second. The constant gain is 0.0625, or

$$K_{dB} = -24.08 \text{ dB}$$

The amplitudes and phase shifts caused by the individual factors are plotted in Figures 6–37(a) and 6–37(b). The Bode plots for the complete system are shown in Figures 6–37(c) and 6–37(d).

The constant-gain term, 0.0625, gives a horizontal line at -24.08 dB on the amplitude plot and no phase shift. The zero at the origin, $j\omega$, yields an amplitude increase of 20 dB per decade, with the response passing through 0 dB at 1 radian per second, or 0.1592 Hz. The zero gives a phase shift of $90°$ at all frequencies.

The amplitude approximations for the poles are constant at 0 dB up to their break or corner frequencies and decrease at 20 dB per decade above those points. The corner frequencies are at the points where the magnitudes of the real and imaginary components of the factor are equal. These are at 2 and 20 radians per second, or 0.3183 and 3.183 Hz, respectively. Figure 6–37(c) shows the sum of the individual amplitude components.

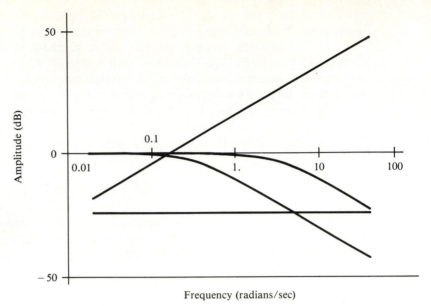

(a) Amplitude versus frequency for individual components

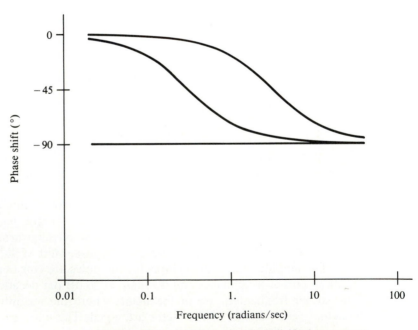

(b) Phase shift versus frequency for individual components

FIGURE 6–37 Bode plot of complex system

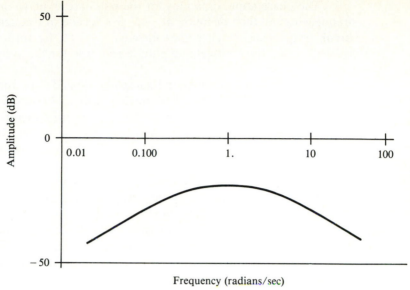

(c) Amplitude versus frequency for total system

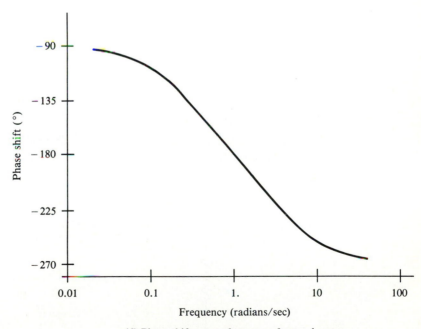

(d) Phase shift versus frequency for total system

FIGURE 6–37 Continued

The phase approximations for the poles are $0°$ up to one-tenth of the corner frequencies and then decrease at $-45°$ per decade of frequency up to 10 times the break frequencies. Each passes through $-45°$ at the corner frequencies. Figure 6–37(d) shows the complete amplitude response for the system.

The technique presented in Examples 6–13 and 6–14 works reasonably well for amplitudes of the simpler transfer function terms and gives an approximation for the phase shifts. If more accuracy is required, the errors incurred by the straight-line approximations are known and corrections may be made. This is most easily done when the break frequencies are far apart. For the preceding example, the errors in one approximation are quite small while the errors in the other are significant, due to the separation in corner frequencies. The exact amplitude response of the transfer function is plotted in Figure 6–37(a). The phase responses overlap much more, which decreases the accuracy of the approximations.

When there is a pair of complex roots, finding the approximate response is more difficult because there are no easily drawn approximations for the responses. One method is to sum the straight-line approximations for the roots on the real axis and then add the response of the complex roots at selected points. Particular care is necessary around the peak of the response for the complex term.

PROBLEMS

6–1. Find the voltage transfer function V_2/V_1 and the transfer impedance V_2/I_1 for the circuit in Figure 6P–1.

6–2. Find the current transfer function I_2/I_1 and the transadmittance I_2/V_1 for the circuit in Figure 6P–2.

6–3. Find the transfer impedance V_2/I_1 for the network in Figure 6P–3.

6–4. Find the transfer admittance I_2/V_1 for the circuit in Figure 6P–4.

6–5. Find the voltage transfer function V_2/V_1 for the circuit shown in Figure 6P–5.

6–6. Find the voltage transfer function V_2/V_1 for the ladder network in Figure 6P–6.

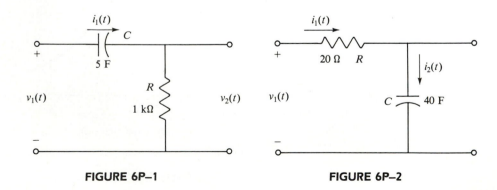

FIGURE 6P–1   **FIGURE 6P–2**

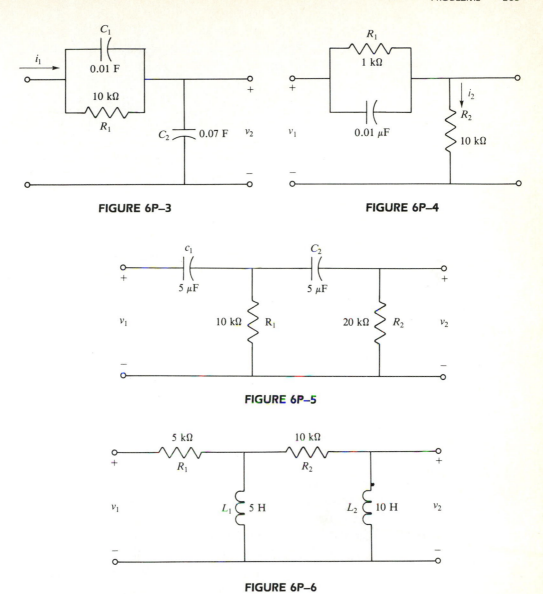

FIGURE 6P–3

FIGURE 6P–4

FIGURE 6P–5

FIGURE 6P–6

6–7. Find the current transfer function I_2/I_1 for the network shown in Figure 6P–7.

6–8. A network with only two terminals has network functions. If the current I into one terminal is the input and the terminal voltage V is the output, then V/I is called the *driving point impedance*. If the voltage is the input and the current is the output, then I/V is the *driving point admittance*. Find the driving point admittance and impedance for the two-terminal network shown in Figure 6P–8.

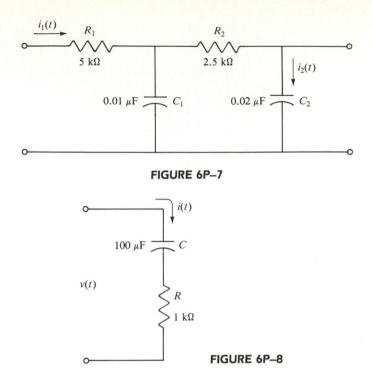

FIGURE 6P–7

FIGURE 6P–8

6–9. Plot the poles and zeros for the functions in Problems 6–1, 6–3, 6–5, and 6–7. Describe the general characteristics of the natural responses.

6–10. Plot the poles and zeros for the functions in Problems 6–2, 6–4, 6–6, and 6–8. What are the general characteristics of the natural responses?

6–11. Write the s-plane mesh equations for the circuit shown in Figure 6P–9. Using any method, find the voltage transfer function V_2/V_1 and the input impedance V_1/I_1.

6–12. What is the voltage transfer function for the AC bridge circuit illustrated in Figure 6P–10? What is its value at balance?

6–13. Figure 6P–11 shows an audio transformer. Using the equivalent circuit described in Chapter 1, find the voltage and current transfer functions. In the transformer, L_1 is 10 H, L_2 is 20 H, and the coupling coefficient k is 0.75. The input impedance is the input voltage divided by the input current, V_1/I_1. What is the value of the input impedance?

6–14. A common equivalent circuit approximation for a transistor uses hybrid parameters. A common emitter amplifier circuit employing hybrid parameters is shown in Figure 6P–12. Find the voltage and current transfer functions.

6–15. Find the impulse and step responses for the circuits in Problems 6–1, 6–5, and 6–7. Give the answer in both the s-plane and the time domain.

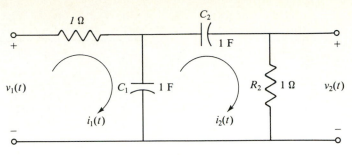

FIGURE 6P-9

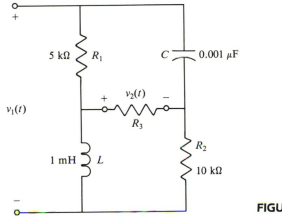

FIGURE 6P-10

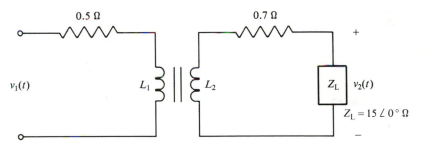

FIGURE 6P-11

6-16. Find the s-plane and time domain impulse and step responses for Problems 6-2, 6-6, and 6-8.

6-17. Sketch the approximate Bode amplitude and phase responses for the circuits whose voltage transfer functions are

a. $T(s) = \dfrac{10}{s + 2}$

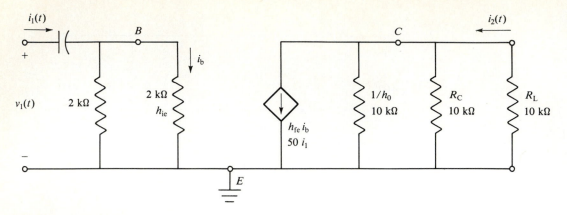

FIGURE 6P–12

    b. $T(s) = \dfrac{30(s + 5)}{s + 15}$

    c. $T(s) = \dfrac{4}{(s + 3)(s + 30)}$

    d. $T(s) = \dfrac{4s}{s^3 + 20s^2 + 100s}$

6–18. Sketch the straight line Bode plot approximations for the systems whose transfer functions are

    a. $T(s) = \dfrac{16s}{s + 8}$

    b. $T(s) = \dfrac{20(s + 30)}{3(s + 20)}$

    c. $T(s) = \dfrac{500s}{s^2 + 120s + 2000}$

    d. $T(s) = \dfrac{3s + 6}{s^3 + 75s^2 + 1000}$

Appendix A

Cramer's Rule

In most cases, the solution of simultaneous equations used in analyzing electric circuits will be performed by the application program that is used to solve the circuits. If the equations are the result of mesh analysis, they will be in the form

$$Z_{11} i_1 + Z_{12} i_2 + \cdots + Z_{1n} i_n = v_1$$
$$Z_{21} i_1 + Z_{22} i_2 + \cdots + Z_{2n} i_n = v_2$$
$$\cdot$$
$$\cdot$$
$$\cdot$$
$$Z_{n1} i_1 + Z_{n2} i_2 + \cdots + Z_{nn} i_n = v_n$$

Cramer's rule says that the solution of this set of equations is

$$i_1 = \frac{D_1}{D}, \ i_2 = \frac{D_2}{D}, \ i_3 = \frac{D_3}{D}, \ldots, i_n = \frac{D_n}{D}$$

where the determinants D, D_1 and D_n are as follows:

$$D = \begin{vmatrix} Z_{11} & Z_{12} & \cdot & \cdot & \cdot & Z_{1n} \\ Z_{21} & Z_{22} & \cdot & \cdot & \cdot & Z_{2n} \\ \cdot & \cdot & \cdot & \cdot & \cdot & \cdot \\ \cdot & \cdot & \cdot & \cdot & \cdot & \cdot \\ \cdot & \cdot & \cdot & \cdot & \cdot & \cdot \\ Z_{n1} & Z_{n2} & \cdot & \cdot & \cdot & Z_{nn} \end{vmatrix}$$

$$D_1 = \begin{vmatrix} v_1 & Z_{12} & \cdot & \cdot & \cdot & Z_{1n} \\ v_2 & Z_{22} & \cdot & \cdot & \cdot & Z_{2n} \\ \cdot & \cdot & & \cdot & \cdot & \cdot \\ \cdot & \cdot & & \cdot & \cdot & \cdot \\ \cdot & \cdot & & \cdot & \cdot & \cdot \\ v_n & Z_{n2} & \cdot & \cdot & \cdot & Z_{nn} \end{vmatrix}$$

$$D_n = \begin{vmatrix} Z_{11} & Z_{12} & \cdot & \cdot & \cdot & v_1 \\ Z_{21} & Z_{22} & \cdot & \cdot & \cdot & v_2 \\ \cdot & \cdot & & \cdot & \cdot & \cdot \\ \cdot & \cdot & & \cdot & \cdot & \cdot \\ \cdot & \cdot & & \cdot & \cdot & \cdot \\ Z_{n1} & Z_{n2} & \cdot & \cdot & \cdot & v_n \end{vmatrix}$$

Determinants can be evaluated by means of the Laplace expansion. The first step is to find the cofactors of each element in a single row or column. For example, for the element in the ith row and jth column of determinant D, the cofactor is

$$C_{ij} = (-1)^{i+j} \begin{vmatrix} Z_{11} & \cdot & \cdot & \cdot & Z_{1,j-1} & Z_{1,j+1} & \cdot & \cdot & \cdot & Z_{1n} \\ Z_{21} & \cdot & \cdot & \cdot & Z_{2,j-1} & Z_{2,j+1} & \cdot & \cdot & \cdot & Z_{2n} \\ \cdot & \cdot & & \cdot & \cdot & \cdot & \cdot & & \cdot \\ \cdot & \cdot & & \cdot & \cdot & \cdot & \cdot & \cdot & & \cdot \\ Z_{i-1,1} & \cdot & \cdot & & \cdot & & \cdot & \cdot & Z_{i-1,n} \\ Z_{i+1,1} & \cdot & \cdot & & \cdot & & \cdot & \cdot & Z_{i+1,n} \\ \cdot & \cdot & & \cdot & \cdot & \cdot & \cdot & & \cdot \\ \cdot & \cdot & & \cdot & \cdot & \cdot & \cdot & \cdot & & \cdot \\ Z_{n,1} & \cdot & \cdot & \cdot & \cdot & \cdot & \cdot & \cdot & \cdot \end{vmatrix}$$

Then the determinants may be evaluated along either the eliminated column or the eliminated row. For example, to evaluate determinant D along the ith row, the equation is

$$D = Z_{11} C_{11} + Z_{i2} C_{i2} + \cdots + Z_{in} C_{in}$$

Along the jth column, we have

$$D = Z_{1j} C_{1j} + Z_{2j} C_{2j} + \cdots + Z_{nj} C_{nj}$$

The signs of the initial multipliers, $(-1)^{i+j}$ can be seen to alternate.

The procedure is much simplified for second- and third-order determinants, as shown in many introductory texts on circuit analysis. For a second-order determinant, the equation is

$$D = \begin{vmatrix} Z_{11} & Z_{12} \\ Z_{21} & Z_{22} \end{vmatrix} = Z_{11}Z_{22} - Z_{21}Z_{12}$$

The value of a third-order determinant is found through the Laplace transformation. We have

$$D = \begin{vmatrix} Z_{11} & Z_{12} & Z_{13} \\ Z_{21} & Z_{22} & Z_{23} \\ Z_{13} & Z_{23} & Z_{33} \end{vmatrix}$$

$$= Z_{11}\begin{vmatrix} Z_{22} & Z_{23} \\ Z_{32} & Z_{33} \end{vmatrix} - Z_{21}\begin{vmatrix} Z_{12} & Z_{13} \\ Z_{32} & Z_{33} \end{vmatrix} + Z_{31}\begin{vmatrix} Z_{12} & Z_{13} \\ Z_{22} & Z_{23} \end{vmatrix}$$

where the second-order matrices are evaluated as just shown.

Derivations of Transforms

The derivations of two of the transform pairs in Table 4–1 were not given in Chapter 4, because of the similarity to other derivations given there. The simple formula for the inversion of transforms of underdamped responses was also not included in that chapter, because the development would likely not help in remembering the formula. For completeness, the transform pairs and the inversion formula will be derived in this appendix.

TRANSFORM OF sin ωt

We can find the Laplace transform of a sine wave with the use of Euler's equation:

$$\mathscr{L}[\sin \omega t] = \mathscr{L}\left|\frac{\epsilon^{j\omega t} - \epsilon^{-j\omega t}}{2j}\right| = \frac{1}{2j}\{[\epsilon^{j\omega t}] - [\epsilon^{-j\omega t}]\}$$

$$= \frac{1}{2j}\left|\frac{1}{s - j\omega} - \frac{1}{s + j\omega}\right| = \frac{\omega}{s^2 + \omega^2}$$

TRANSFORM OF THE RAMP FUNCTION

The Laplace transform of the ramp is given by

$$\mathscr{L}[t] = \int_0^\infty t\epsilon^{-st}\, dt$$

The integral can be evaluated using integration by parts:

$$\int u\, dv = uv - \int v\, du$$

Letting

$$u = t \qquad du = dt$$

$$dv = \epsilon^{-st} \qquad v = \frac{\epsilon^{-st}}{-s}$$

and substituting, we get

$$\mathcal{L}[t] = \left| \frac{t\epsilon^{-st}}{-s} - \int \frac{\epsilon^{-st}}{-s} \, dt \right|_0^\infty = \frac{1}{s^2}$$

FORMULA FOR INVERSION WHEN THERE IS A PAIR OF COMPLEX ROOTS

If an s-plane response $F(s)$ has a pair of complex poles, the product of the complex roots can be put into the form

$$(s + a)^2 + \omega^2$$

If the complex roots are factored out of $F(s)$, we have

$$F(s) = \frac{Q(s)}{(s + a)^2 + \omega^2}$$

The factors of the quadratic term are $s + a - j\omega$ and $s + a + j\omega$. Partial fraction expansion of $F(s)$ gives

$$F(s) = \frac{K_1}{s + a - j\omega} + \frac{K_2}{s + a + j\omega} + R(s)$$

The first two terms are due to the complex roots, and $R(s)$ includes all other terms. From Chapter 4, K_2 is the complex conjugate of K_1, or

$$K_2 = K_1^*$$

The partial fraction expansion procedure can be used to find the numerators, K_1 and K_1^*. We have

$$K_1 = \frac{Q(s)}{s + a - j\omega} \bigg|_{s = -a + j\omega} = \frac{Q(-a - j\omega)}{-2j\omega} = \frac{A/\theta}{2j\omega} = \frac{A\epsilon^{j\theta}}{2j\omega}$$

and

$$K_1^* = \frac{Q(s)}{s + a + j\omega}\bigg|_{s = -a - j\omega} = \frac{Q(-a - j\omega)}{-2j\omega} = \frac{A\angle\theta}{-2j\omega} = \frac{Ae^{j\theta}}{-2j\omega}$$

Substituting these values into the expression for $F(s)$, we get

$$F(s) = \frac{Ae^{j\theta}}{2j\omega(s + a - j\omega)} - \frac{Ae^{-j\theta}}{2j\omega(s + a + j\omega)} + R(s)$$

The inverse transform of $F(s)$ is

$$f(t) = \left[\frac{Ae^{j(\omega t + \theta)} - Ae^{-j(\omega t + \theta)}}{2j\omega}\right]e^{-at} + r(t)$$

where $r(t)$ is the inverse transform of $R(s)$. Simplifying the first term using Euler's equation, we obtain

$$f(t) = \frac{A}{\omega} e^{-at} \sin(\omega t + \theta) + r(t)$$

Appendix C

SPICE

C.1 INTRODUCTION

SPICE is an acronym for Simulation Program with Integrated Circuit Emphasis. The original program was developed for use on mainframe computers in the mid-1970s at the University of California at Berkeley. Versions have been developed for use on minicomputers and microcomputers since then. As a result, SPICE has effectively become the standard circuit analysis program.

One microcomputer-based version, called PSpice, has been developed for use on personal computers, workstations, and minicomputers and was used to check the results of examples used in this text. While still following the Berkeley SPICE2 standards, PSpice has been overhauled to avoid some residual errors in the original. It also has a number of additions and improvements. One of the improvements is the PROBE program, a graphical curve-drawing program that augments the printer plot output available from the basic PSpice program.

SPICE programs have a number of tools that are not applicable to passive circuit analysis. We will therefore not include instruction in how to use those tools. The PROBE graphics curve-plotting program is useful, but it can be easily used without instruction. All that needs to be done is to include the .PROBE directive as a part of the circuit file. The program has a menu system that is fairly easy to use along with trial-and-error methods. Care must be taken to make sure that the data are sampled often enough so that the curve-drawing routine properly represents the actual curve. Errors are likely to occur at points where the slope changes rapidly, such as peaks of sine waves that have a number of cycles per inch.

Some limited experience with a demonstration disk for another microcomputer-based version of SPICE called IsSpice indicates that this version is generally similar in operation to PSpice. Its graphics processor is called IntuScope.

This appendix describes how to use SPICE to solve the problems and examples in Chapters 5 and 6 of this book. It is not intended to cover all the features of the program. For more information about the use of SPICE programs, consult any of the documents listed at the end of this appendix.

C.2 TYPES OF ANALYSIS

The analytical procedures used with SPICE to prepare examples in this text are the frequency response and transient response analyses. Transient response analysis was used in Chapter 5, and frequency response analysis was used in Chapter 6. SPICE programs calculate the DC bias point for all circuits before proceeding to other types of analysis. This technique may be used to analyze a DC circuit. A number of other types of analysis not used in this text are also available.

C.3 GENERAL PROCEDURE

The first step in the use of SPICE is to sketch the circuit and number each node or junction between components. Each component and source is given an alphanumeric name. The first character is a letter indicating the type of component. For example, R5 is a resistor and VIN is an input voltage. Then the inpt file is developed. Figure C–1 shows a simple voltage divider circuit. The input, or circuit, file for this circuit is as follows:

```
Example
* Voltage divider
VIN 1 0 5
R1   1 2 5
R2   2 0 10
.END
```

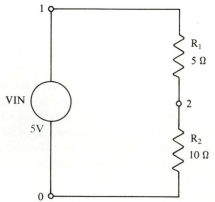

FIGURE C–1 Voltage divider

This file is incomplete because it has no line or lines describing the analysis to be performed. In a complete circuit file, such lines would often be between the lines describing the circuit and the end statement. The first line is the title. This will be printed on the top line of both the circuit file and the output file listings. It does not have to be the file name, but it could be. If so, the file shown would be called EXAMPLE.CIR. The second line, since it starts with an asterisk, is a comment. The component and source lines follow. The syntax is

⟨name⟩ ⟨+node⟩ ⟨−node⟩...⟨value⟩

The simpler elements will have only two nodes, but multiterminal devices will obviously have more. The lines following the circuit listing are called *directives* and are preceded by a period. Among other things, they determine the type of analysis to be performed and the desired outputs. The end statement in the circuit listing is a directive.

When the circuit file EXAMPLE.CIR is analyzed, the output file name is EXAMPLE.OUT. The types of analysis described in this appendix are AC and transient analysis, for passive circuits. Other types of analysis are possible with SPICE, but they will not be covered here.

C.4 DECIMAL MULTIPLIERS

In expressing powers of 10, scientific notation, expressed as a floating-point number, can be used. That is, 100 may be written as 1E2. It is also possible to use prefix letters, much like those used in electronic circuit diagrams. The specific letters, along with their equivalent powers of 10, are as follows:

F	10^{-15}
P	10^{-12}
N	10^{-9}
U	10^{-6}
M	10^{-3}
K	10^{3}
MEG	10^{6}
G	10^{9}
T	10^{12}

The case of the prefixes (upper or lower) is not important, and the actual units—volts, amperes, or ohms—do not need to be included. The units of an inductor are henrys, those of a capacitor are farads, and those of a resistor are ohms. The program will ignore any other letters once it identifies the prefixes. The two systems may be used together, as long as the prefixes follow the E and its associated numerals. That is, 1.056E9, 1.056MEG, and 1.056E6K are all equal.

C.5 PASSIVE COMPONENTS

The simplest linear passive components used in electric circuits are resistances, capacitors, and inductors. Other more complex components can be represented by equivalent circuits consisting of the same three components. Mutual inductance may be specified by its coefficient of coupling. In SPICE programs, these components are named by using the letter commonly used to identify them in circuit analysis:

resistor	Rxx
capacitor	Cxx
self-inductance	Lxx
coupling coefficient	Kxx

The xx stands for a unique set of alphanumeric characters that identify the specific component or parameter. In some older versions of SPICE, the names must be no more than eight character long.

C.5.1 Specification of Components

To specify the components completely, we can list them by their names in the following fashion:

```
R〈name〉 〈node〉 〈node〉 〈value〉
C〈name〉 〈node〉 〈node〉 〈value〉 [IC = xx]
L〈name〉 〈node〉 〈node〉 〈value〉 [IC = xx]
```

The first node is the one that positive current flows into. The items within angle brackets are required, and those within square brackets are optional. The initial conditions (IC) are the initial voltage for capacitors and the initial current for inductors. These are used only in transient analyses and require a special term in the statement calling up the transient analysis. The default values are zero. Typical component specifications are as follows:

RIN 2 3 1MEG: a 1-MΩ resistor, RIN, between nodes 2 and 3.

C2 3 4 20U IC=5: a 20-μF capacitor, C2, between nodes 3 and 4, with an initial voltage of 5 V.

L33 4 0 5M IC=3Ma.: a 5-mH inductor, L33, between nodes 4 and 0, with an initial current of 3 mA.

K12 L1 L2 0.79: the coefficient of coupling between inductors L1 and L2, with value 0.79.

The components specified in the foregoing manner are linear elements.

C.5.2 Special Considerations for Components

In DC circuits, no current will flow through capacitors under steady-state conditions As a result, if any capacitor has no path to ground, bias calculations in DC analysi cannot be made. A very large resistor may be connected between an isolated node and ground to prevent this condition without any significant effect on the operation of the circuit. The resistor can simulate the leakage resistance that would be present in a real capacitor.

An ideal inductor, such as those in the SPICE models, has no resistance. If it is connected directly across an ideal source or another inductor, it forms what is called a voltage loop. Voltage loops must be prevented by connecting a resistor with low resistance in the loop.

C.6   ACTIVE DEVICES

Active devices, such as transistors, are not needed to solve the circuit problems found in this text. If needed for other purposes, ideal and real active devices may be simulated using the ideal controlled-source simulations provided by the programs. Simulations of real active devices and circuits are available as library circuits with PSpice and other versions of SPICE. More information on their use is available in the literature listed at the end of this appendix.

C.7   INDEPENDENT SOURCES

Ideal independent voltage and current sources can be identified in a manner similar to that used for passive components. That is, Vxx is a voltage source and Iyy is a current source, where xx and yy are used to identify the specific sources. Real voltage and current sources may be simulated by connecting the correct value in series with the voltage or current sources.

For sources, it is necessary to specify which terminal is connected to each node in order to provide the proper polarity or direction of current flow. A source may have DC, AC, and/or transient components. The default value for any component is zero. If the type of source is not specified, the default is DC.

C.7.1   Voltage sources

The general forms for voltage sources are

```
V⟨name⟩ ⟨+node⟩ ⟨−node⟩ [[DC] ⟨value⟩]
V⟨name⟩ ⟨+node⟩ ⟨−node⟩ AC ⟨magnitude value⟩ [phase value]
V⟨name⟩ ⟨+node⟩ ⟨−node⟩ (transient specification)
```

$$(c-1)$$

for DC, AC, and transient sources, respectively. Items within angle brackets are required, and those within square brackets are optional. A source may have components of all three types. If so, the voltage specifications may be combined following the name and node designations. If more than one line is required for any source description, the continuation lines should start with a plus sign.

A 15-kV DC source called VIN with its positive terminal connected to node 1 and its negative terminal connected to node 2 is written as

```
VIN 1 2 15K
```

An AC source of 65 volts amplitude with a phase angle of 35° would be given by

```
VS 3 4 65 35
```

The frequency of the source would be defined in an AC control statement. This specification will be shown in the section on control statements, as will some of the possible transient specifications. To print or plot the foregoing voltages, we use the output specifications, VIN and VS.

C.7.2   Current Sources

The general forms for current sources are similar to those for voltage sources:

```
I⟨name⟩ ⟨+node⟩ ⟨−node⟩ [[DC] ⟨value⟩]
I⟨name⟩ ⟨+node⟩ ⟨−node⟩ AC ⟨magnitude⟩ [phase]          (c−2)
I⟨name⟩ ⟨+node⟩ ⟨−node⟩ (transient specification)
```

Conventional current will flow into the positive node through the current source. Current sources can provide DC, AC, or transient currents.

C.7.3   Measurement of Current

If it is necessay to find the current at some point in a circuit, a zero-volt voltage source is placed in the circuit so that the current flows through it. This zero-volt voltage source is specified by

```
V⟨name⟩ ⟨+node⟩ ⟨−node⟩ [DC] 0
```

The current is then found by the output statement, I(V⟨name⟩).

C.8   TRANSIENT ANALYSES

Transient responses can be analyzed by including the control statement .TRAN. This statement specifies the length of the response desired, the desired printer output range (for both tables and printer plots), and the maximum time step. The syntax is

```
.TRAN ⟨print step interval⟩ ⟨final time⟩
+[no-print interval [maximum step time]] [UIC]
```

The print step interval controls the time between printed data points, both for tabular output and printer plots. The length of the calculated response is specified by the final time. The no-print interval will prevent any printer output from appearing for the stated time. If that interval is included, the maximum calculation step interval may be also listed. The default step ceiling is the final time divided by 50. A value larger or smaller than that may be specified, however.

The UIC statements causes the initial conditions stated in the component listings (IC=xx) to be used. The control statement, IC, can also be used to define initial condition voltages at circuit nodes.

In a transient analysis, sources are defined in the manner described in equations (C–1) and (C–2). AC sources are set to zero, but DC and transient sources are used in the analysis. The transient specifications include exponential pulses (EXP), quadrilateral pulses with flat tops and bottoms that can be repetitive (PULSE), piecewise linear waveforms (PWL), single-tone frequency-modulated signals (SFFM), and switched sinusoids (SIN), which can be multiplied by exponentials. These transient signals will have zero amplitudes in steady-state analyses.

The general format for the EXP specification is

```
EXP (⟨v1⟩ ⟨v2⟩ ⟨td1⟩ ⟨tc1⟩ ⟨td2⟩ ⟨tc2⟩)
```

In this specification, v1 is the initial voltage, v2 is the peak voltage, td1 is the delay of the initial rise with a default value of 0 sec, and tc1 is the rise time constant with a default of the transient print step (TSTEP). For the fall of the pulse, td2 is the delay of the fall with a default of td1 plus TSTEP, and tc2 is the fall time constant, whose value is TSTEP. Figure C–2 shows an exponential pulse with the specification

```
V1 1 0 EXP(1v 5v 1 0.2 2 0.5)
```

The pulse transient specification, which can produce single or repetitive pulses, is

```
PULSE(⟨v1⟩ ⟨v2⟩ ⟨td⟩ ⟨tr⟩ ⟨tf⟩ ⟨pw⟩ ⟨per⟩)
```

Here, v 1 is the initial voltage and v2 is the voltage of the flat pulse top. There is no default for these voltages. The time delay td is the delay before the start of the first rise with a zero default. The rise time and fall time are tr and tf, respectively. Their default is TSTEP. The pulse width is pw, and its default is TSTOP, the ending time for the transient analysis. The period for repetitive pulses is per. Its default value is also TSTOP. Figure C–3 is a flat sided pulse with the specification

```
V2 2 0 PULSE(1v 5v 0 0.2 0.3 0.5 2)
```

The pulses can be seen to have flat tops and bottoms and to be constructed of straight lines.

The piecewise linear signal specification produces a signal constructed by

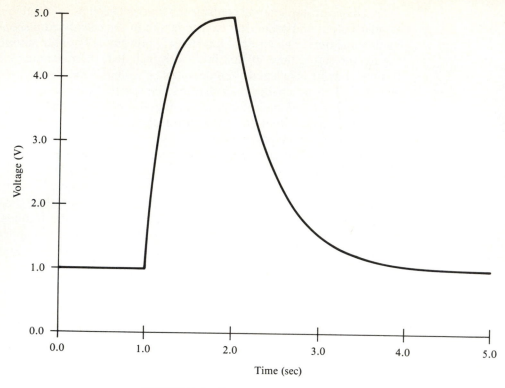

FIGURE C–2   Exponential transient source

straight lines joining a sequence of points on the plot of voltage versus time. Its general specification is

```
PWL(⟨t1⟩ ⟨v1⟩ ⟨t2⟩ ⟨v2⟩ ..... ⟨tn⟩ ⟨vn⟩)
```

The value tn is the time at the nth corner, and vn is the voltage. There are no defaults. Figure C–4 is a piecewise linear waveform with the specification

```
V3 3 0 PWL(0 1V 1 1V 1.2 5V 1.5 4V 2 2V 3 3V)
```

The single-frequency frequency-modulated transient signal has the general specification

```
SFFM(⟨voff⟩ ⟨vampl⟩ ⟨fc⟩ ⟨mod⟩ ⟨fm⟩)
```

The voltage parameters are voff, the offset voltage, and vampl, the peak amplitude. They have no default values. The carrier frequency is fc and has a default value of 1/TSTOP. The modulation index is mod and has a default of zero. The modulation

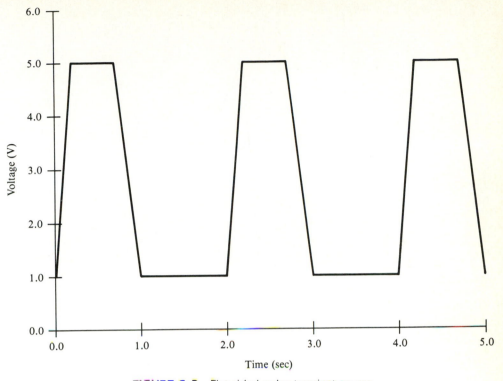

FIGURE C–3 Flat-sided pulse transient source

frequency is fm, with a default value of 1/TSTOP. Figure C–5 shows a modulated signal with the specification

```
V4 4 0 SFFM((2V 1V 8Hz 4 1Hz)
```

The sinusoid transient signal can have an exponential multiplier. Its general specification is

```
SIN(〈voff〉 〈ampl〉 〈freq〉 〈td〉 〈df〉 〈phase〉)
```

The offset voltage is voff, and the peak amplitude is vampl. Neither of these has a default value. The frequency is freq, with a default value of 1/TSTOP. Its starting time is td, and its default is zero. The damping factor is df. If df is greater than zero, the sine wave decays, but if it is negative, the sinusoid increases exponentially. Figure C–6 shows a single-tone frequency-modulated modulation signal with specification

```
V5 5 0 SIN(2 2 5HZ 1 1 30)
```

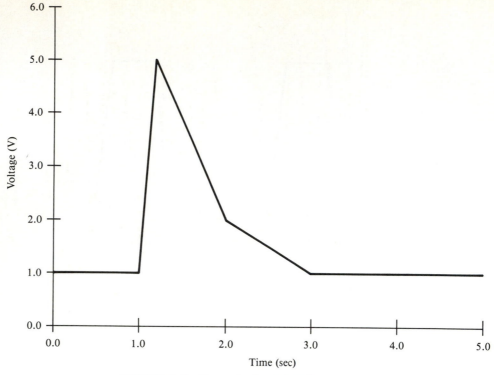

FIGURE C–4   Piecewise linear transient source

C.9   BODE PLOTS

Bode plots are generated by specifying the circuit as with any other circuit. The source is specified as a 1-V AC source with a default phase shift of 0°. That is,

```
VIN 1 0 AC 1
```

would be an AC source of amplitude 1 V with its positive terminal numbered 1 and its negative terminal numbered 0. An AC analysis command must also be included. The general form of this command is

```
.AC [LIN][OCT][DEC] ⟨(points) value⟩
+⟨(start frequency) value⟩ ⟨(end frequency) value⟩
```

LIN signifies a linear sweep; that is, the frequency is swept linearly from the starting to the ending frequency. The total number of points in the sweep is specified by ⟨(points) value⟩. OCT denotes a logarithmic sweep by octaves. The number of points per octave is specified by ⟨(points) value⟩. DEC signifies a logarithmic sweep

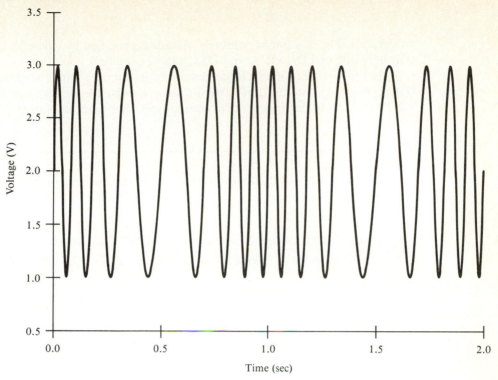

FIGURE C–5   FM transient source

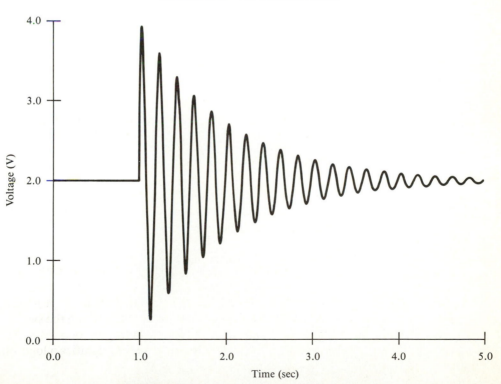

FIGURE C–6   Damped sine wave transient source

by decades. The number of points per decade is specified by ⟨(points) value⟩. Note that only one of LIN, OCT, and DEC can be specified.

In the specification of a Bode plot, ⟨(end frequency) value⟩ must be lower than ⟨(start frequency) value⟩. If both are the same, a single frequency response is found. All DC and transient independent sources will have zero amplitude.

C.10   REFERENCES

In addition to the documentation for commercial versions of SPICE, there are several books that can be used to find out more about its use. Recent publications include:

- Hines, John R. *A Spice Tutorial.* Richardson, TX: Ohiolab Technology, 1988.
- Hines, John R. *A Spice Users Guide.* Richardson, TX: Ohiolab Technology, 1988.
- Tuinenga, Paul W. *SPICE: A Guide to Circuit Simulation & Analysis Using PSpice.* Englewood Cliffs, NJ: Prentice Hall, 1988.

Answers to Selected Problems

Chapter 1

1–1. $v_i = 10$ V, $i_R = 5$ A, $i_e = 0$ A

1–3. $i = 5\angle 120°$ A, $v = \angle -18°$ V

1–5. $28.27\angle 108°$ mA

1–7. $82.47 \sin (350\pi t + 80°)$ V

1–9. $174.4\angle -125.0°$ Ω, $21.22\angle -90°$ Ω

1–11. $e_1 = 7.5i_1 + 10\dfrac{di_1}{dt} - 3.098\dfrac{di_2}{dt}$

$0 = (6.5 + j4)i_2 + 5\dfrac{di_2}{dt} - 3.098\dfrac{di_1}{dt}$

1–13. $I_1 = -90.91$ mA, $I_2 = -454.5$ mA

1–15. $V_1 = 327.8\angle 40.10°$ V,
$V_2 = 369.7\angle 63.80°$ V

1–17. 2.052 A

1–19. $V_{OC} = V_{Th} = 24$ V, $I_{sc} = I_N = 2.4$ A,
$Z_{Th} = Z_N = 10$ Ω, $V_L = 4$ V, 6.857 V,
9.0 V

Chapter 2

2–3. **a.** $5.000 \sin (0.4488t + 77.14°)$ V
b. $0.700 \sin (0.5236t - 60°)$ A

2–5. **a.** $45.24 \sin (120\pi t + 90°)$ mV/sec
b. $-32.67 \sin (800\pi t + 55°)$ mA/sec

2–7. **a.** $0.2000 \sin 90t$ A-sec
b. $5.429 \sin (35t - 41°)$ V-sec

2–11. $v(t) = 7u(t - 5) - 7u(t - 10)$ V,
$v(t) = 7u(t - 5) + 3u(t - 7)$
$\quad - 10u(t - 10)$ V,
$v(t) = 5u(t - 2) - 10u(t - 7)$
$\quad + 5u(t - 12)$ V

2–13. $v(t) = 1.2tu(t) - 2.4(t - 10)u(t - 10)$
$\quad + 1.2(t - 20)$ V,
$v(t) = 2.4(t - 5)u(t - 5)$
$\quad - 2.4(t - 10)u(t - 10)$
$\quad - 2.4(t - 20)u(t - 20)$
$\quad + 2.4(t - 25)u(t - 25)$ V,
$i(t) = 1.4tu(t) - 1.4(t - 5)u(t - 5)$
$\quad - 1.4(t - 10)u(t - 10)$
$\quad + 1.4(t - 20)u(t - 20)$
$\quad + 1.4(t - 25)u(t - 25)$
$\quad - 1.4(t - 30)u(t - 30)$ A

2–15. $v(t) = 15 \exp \dfrac{-(t - 10)}{1.5} u(t - 10)$

$\quad - 15 \exp \dfrac{-(t - 12)}{1.5} u(t - 12)$

$\quad - 10 \exp \dfrac{-(t - 12)}{1.5} u(t - 12)$

$\quad - 10 \exp \dfrac{-(t - 14)}{1.5} u(t - 14)$ V

Chapter 3

3–1. $\dfrac{di}{dt} + 10^6 i = 10^7$

3–3. $\dfrac{dv}{dt} + 200v = 0$

3–5. $\dfrac{d^2 v}{dt^2} + 2000\dfrac{dv}{dt} + 25 \times 10^6 v = 0$

3–7. $\dfrac{d_i^2}{dt^2} + 2 \times 10_i^4 = 0$

3–9. $s + 10^6 = 0$

3–11. $s^2 + 2000s + 25 \times 10^6 = 0$

3–13. $i_f(t) = 0$ A, $i_n(t) = i(t) = 0.05\epsilon^{-6250t}$ A,
$v_R(t) = 5^{-6250t}$ V, $v_C(t) = 5 - 5\epsilon^{-6250t}$ V

3–15. $f(t) = M\dfrac{dv}{dt} + Bv$, $i(t) = C\dfrac{dv}{dt} + \dfrac{v}{R}$

3–17. $K\!\int (v_1 - v_2)\,dt + M\dfrac{dv_1}{dt} + Bv_1 = 0$,
$f(t) = k\!\int (v_2 - v_1)\,dt$

Chapter 4

4–1. $\dfrac{5}{s}$

4–3. $\dfrac{15s}{s^2 + 1225}$

4–5. $\dfrac{875}{s^2 + 1225}$

4–7. $\dfrac{35}{s^2}$

4–9. $\dfrac{37.5(s + 22)}{s^2 + 44s + 2509}$

4–11. 32

4–13. $\dfrac{5s + 8}{s^2}$

4–15. $\dfrac{7s - 8}{s^2}$

4–17. $\dfrac{5}{s^2}$

4–19. $115\epsilon^{-5s}$

4–21. $\dfrac{45\epsilon^{-s}}{s^2}$

4–23. $\dfrac{5\epsilon^{-s}}{s} - \dfrac{5\epsilon^{-3s}}{s}$

4–25. $\dfrac{5}{s} + \dfrac{2}{s^2} - \dfrac{4\epsilon^{-5s}}{s} + \dfrac{2\epsilon^{-12.5s}}{s}$

4–27. $-4, -4$

4–29. $\pm j3$

4–31. $-\dfrac{40}{s + 8} + \dfrac{45}{s + 9}$

4–33. $\dfrac{j5}{6(s - 2 + j3)} - \dfrac{j5}{6(s - 2 - j3)}$

4–35. $\dfrac{1}{s + 6} - \dfrac{3}{(s + 6)^2}$

4–37. $\delta(t) + 3.5\epsilon^{-2t} - 11.5\epsilon^{-6t}$

4–39. $10\delta(t) - 15u(t)$

4–41. $\dfrac{\epsilon^{-35t}}{35}$

4–43. $2u(t) - 3.333\epsilon^{-2t} + 1.333\epsilon^{-5t}$

4–45. $2\delta(t) + 4u(t) + 5\epsilon^{-t}$

4–47. $5 \sin (6t + 60°)$

4–49. $0.1\epsilon^{-t} \sin (2t - 126.9°)$
$+ 0.2t\epsilon^{-2t} - 0.08\epsilon^{-2t}$

4–51. $f(0) = 5, f(\infty) = 0; f(0) = f(\infty) = 0;$
$f(0) = 1, f(\infty) = 0$

Chapter 5

5–1. a. $\dfrac{5}{s}$ **b.** $\dfrac{44.83s}{s^2 + (24\pi)^2} + \dfrac{0.8716(24\pi)}{s^2 + (24\pi)^2}$

c. $\dfrac{25}{s + 45}$ **d.** $\dfrac{6.106}{s^2 + 19990} + \dfrac{4440}{s^2 + 19990}$

5–3. a. 35   **b.** $\dfrac{10^5}{s}$ **c.** 30×10^{-3s}

5–5. a. $\dfrac{10^6}{75s} + \dfrac{10}{s}$ **b.** $0.1s + 300 \times 10^{-6}$

c. 10   **d.** $\dfrac{10^{12}}{25s}$

5–7. Thévenin: $Z(s) = \dfrac{10^{12}}{150s}$, $V(s) = \dfrac{10}{s}$

Norton: $Z(s) = \dfrac{10^{12}}{150s}$, $I(s) = 1.5 \times 10^{-9}$

5–9. $E = \dfrac{80}{s}$, $Z_c = \dfrac{10^5}{s}$, $Z_R = 45$

5–11. $I = \dfrac{30 \times 10^{-3}}{s}$, $R_s = 10 \times 10^3$,

$C = \dfrac{2 \times 10^8}{s}$, $R = 20$, $L = 10 \times 10^{-3}s$

5–13. $I = \dfrac{5.736s + 1.287 \times 10^6}{s^2 + 24680}$,

$R_s = 35 \times 10^3$, $C = \dfrac{10^{12}}{75s}$, $R = 82 \times 10^3$,

$L = 10s$, $V_c(0+) = \dfrac{10}{s}$

5–15. $E = \dfrac{6.075s + 324.9 \times 10^3}{s^2 + 88.83 \times 10^6}$,

$R_s = 35 \times 10^3$, $L = 10s$, $C = \dfrac{10^5}{s}$,

$R = 32$, $V_L(0^+) = -3.657 \times 10^3$,
in series.

5–17. a. $10\epsilon^{-5t}$
b. $39 \cos (3.9 \times 10^3 t) + \sin (3.9 \times 10^3 t,$
c. $2 \sin 6t$ **d.** $35\epsilon^{-8t} \cos 5t$

5–19. $0.7692\epsilon^{-1.3t} - 0.7692$ A

5–21. $1.266 \sin (5t + 103.7°) - 132.8\epsilon^{-17.83t}$
$+ 132.8\epsilon - 0.175t(t)$

5–23. $0.37 \times 10^{-3}\delta(t) + 10^{-3}\epsilon^{-2t}$ A

5–25. $50\epsilon^{-2t}$ V, $2\epsilon^{-2t} - 2\epsilon^{-7t}$ A,
$-20\epsilon^{-2t} + 70\epsilon^{-7t}$ V, $70\epsilon^{-2t} - 70\epsilon^{-7t}$ V

5–27. $1.633\epsilon^{-0.5t} \sin (0.3873t + 37.76°)$ A

5–29. $1.026 \times 10^{-3}\epsilon^{-100t} \sin (435.9t - 77.08°)$ A

5–31. $-10.46\epsilon^{-0.9583t} + 0.4554\epsilon^{-0.04174t}$ A

5–33. $E_1 = \dfrac{6}{s}$, $R_1 = 2$, $L = 4s$, $C_1 = \dfrac{10^6}{s}$,

$V_{C1} = \dfrac{14}{s}$, $E_2 = \dfrac{20}{s}$, $V_L = 12$, $R_2 = 3$,

$C_2 = \dfrac{5 \times 10^5}{s}$, $V_{C2} = 20$, $I_2 = \dfrac{30}{s^2 + 9}$

Chapter 6

6–1. $\dfrac{V_2}{V_1} = \dfrac{s}{s + 0.2 \times 10^{-3}}$, $\dfrac{V_2}{I_1} = 1000$

6–3. $\dfrac{14.29}{s}$

6–5. $\dfrac{s^2}{s^2 + 40s + 200}$

6–7. $\dfrac{40 \times 10^3}{s + 60 \times 10^3}$

6–9. (6–1): zero at $s = 0$, pole at
$s = -0.2 \times 10^{-3}$, exponential
decay
(6–3): pole at $s = 0$, step function
(6–5): 2 zeros at $s = 0$, poles at
$s = -5.860, -34.14$, sum of two
exponential decays
(6–7): pole at $s = -60 \times 10^3$,
exponential decay

6–11. $V_1(s) = I_1\left(R_1 + \dfrac{1}{sC_1}\right) - I_2\dfrac{1}{sC_1}$

$0 = -I_1\dfrac{1}{sC_1} + I_2\dfrac{1}{sC_1} + \left(\dfrac{1}{sC_2} + R_2\right)$

$\dfrac{V_2}{V_1} = \dfrac{s}{s^2 + 3s + 1}$, $\dfrac{V_1}{I_1} = \dfrac{s^2 + 3s + 1}{s(s + 2)}$

6–13. $\dfrac{V_2}{V_1} = -\dfrac{1.256}{s^2 + 0.2723s + 0.03926}$,

$\dfrac{I_2}{I_1} = -\dfrac{0.5305s}{s + 0.7850}$

$\dfrac{V_1}{I_1} = \dfrac{10(s + 0.2571)(s + 0.01530)}{s + 0.7850}$

6–15. (6–1): $G(s) = \dfrac{s}{s + 0.2 \times 10^{-3}}$
$g(t) = \delta(t) - 0.2 \times 10^{-3}$
$\exp (-0.2 \times 10^{-3t})$
$H(s) = \dfrac{1}{s + 0.2 \times 10^{-3}}$,
$h(t) = \exp (-0.2 \times 10^{-3})$

(6–5): $G(s) = 1 - \dfrac{39.15}{s + 34.14} - \dfrac{10.86}{s + 5.86}$

$g(t) = \delta(t) - 39.15\epsilon^{-34.14t}$
$\qquad - 10.86\epsilon^{-5.86t}$

$H(s) = \dfrac{1.207}{s + 34.14} - \dfrac{0.2072}{s + 5.860}$

$h(t) = 1.207\epsilon^{-34.14t}$
$\qquad - 0.2072\epsilon^{-5.860t}$

(6–7): $G(s) = \dfrac{20 \times 10^3}{s + 20 \times 10^3}$

$g(t) = 20 \times 10^{-3}$
$\qquad \exp(-20 \times 10^3 t)$

$H(s) = \dfrac{1}{s} - \dfrac{1}{s + 20 \times 10^3}$

$h(t) = 1 - \exp(-20 \times 10^3 t)$

Index